S0-BVP-228

Municipal Solid Waste Collection

Municipal Solid Waste Collection

A State-of-the-Art Study

National Center for
Resource Recovery, Inc.

Lexington Books
D.C. Heath and Company
Lexington, Massachusetts
Toronto London

Library of Congress Cataloging in Publication Data

National Center for Resource Recovery, Inc.
Municipal solid waste collection.

Bibliography: p.
1. Refuse and refuse disposal. I. Title.
TD791.N25 628'.442 73-7550
ISBN 0-669-89433-8

Published simultaneously in Canada.

Printed in the United States of America.

International Standard Book Number:0-669-89433-8

Library of Congress Catalog Card Number: 73-7550

Table of Contents

List of Figures

List of Tables

Foreword

This book is a current state-of-the-art review of the collection functions performed as part of municipally oriented solid waste management activities.

A broad audience is addressed that ranges from the public official to the concerned citizen. Particularly, the objective of this study is to give insight to state and local government officials and business sector participants in solid waste collection functions. It is anticipated, therefore, that this book will be used by public works personnel in administrative and operations positions and by public sector policy making bodies. To fulfill this objective, attention has been focused upon (1) the qualitative and quantitative nature of municipal refuse and its constituent elements from various sites of generation, (2) the problems, equipments, and systems available to solve or ameliorate on-site waste handling and storage conditions, (3) the review of available technology to make the refuse collection and transport function more effective, and (4) the definition and discussion of the existent and conceptual pre-disposal tasks and methodologies that improve and augment the management of municipal refuse.

A second objective has been to satisfy and inform concerned citizens on the nature, extent, and accurate dimensions of waste generation, handling, storage, collection, transportation, and processing functions. Such data may alleviate misunderstandings about the nation's "third pollution" problem and arm people with information for utilization in their communities. A perspective is conveyed of the interrelated problems regarding solid waste character, equipments used in homes and throughout the community, of capital and operations costs of all aspects of the collection function, of new techniques and novel approaches that are conceived for employment in waste management, and of the advantages and disadvantages known or potentially existent in prosecution of refuse collection functions.

Preface

A state-of-the-art presentation regarding solid waste management and the collection of refuse cannot be complete and all-encompassing. This book addresses many points briefly and portrays the essence of the collection function as it exists today. It is not to be considered a superficial treatment of the subject, as attention is paid to definition of the collection functions, solid refuse character and nature, activities and events that occur at the site of generation, the important facets of actual collection and transportation of refuse from the home or business or industrial plant to the place for disposal, and the many new approaches and procedures employed to process wastes prior to its disposal. In addition, guidelines and references are presented for further study on specific points of interest.

Chapter 1 defines the elements inherent in the collection function and shows how these elements interrelate.

In Chapter 2, attention has been directed to a discussion of domestic and nondomestic wastes from the viewpoints of quality and quantity at generation. Special focus has been directed to the wastes that appear in public places.

The storage and handling problems and equipments of residential, business, and public place refuse are discussed in Chapter 3. New concepts employed for waste storage and handling on-site are described, including the vacuum systems and other preprocessing techniques being utilized or undergoing experimentation at this time.

Chapter 4 delves into the personnel aspects of hiring, training, health, safety, and efficiency of the public and private sector organizations responsible for the prosecution of this part of the waste management function. The important features of the financing and the costs for the conduct of the transportation function are described and there is a review of the numerous equipments utilized. To demonstrate the success of modern waste transportation and collection technology, several examples are given of the innovative technology that is now employed to improve these procedures.

In Chapter 5, the subjects of rail and barge haul and transfer station operation are addressed. Other viable concepts used in the processing of wastes prior to disposal are briefly covered.

Chapter 6 presents an extensive bibliography for additional reading on the subject of solid waste collection.

Acknowledgments

The National Center wishes to acknowledge the invaluable participation of Mr. Barry Ashby in the preparation of this study. Appreciation is also extended for the contributions of and advice from the following reviewers: Mr. Leo Goldstein, Professional Engineer; Mr. David H. Marks, Associate Professor, Massachusetts Institute of Technology; Mr. Richard H. Sullivan, Assistant Executive Director, American Public Works Association; Mr. Ward L. Rothgeb, Director of Public Works, City of Rockville, Maryland; and Mr. Ron Vancil, Economic Consultant, The National Center for Resource Recovery, Inc.

1 Collection Functions

From the point in time and place where a bit or piece of solid waste is generated to the point in time and place where that bit of waste is disposed to a final resting place, any number of things may occur which effect that scrap of throw-away. It serves well to illustrate what may occur in the chain of existence of a bit of waste. By the following example, we can define the collection functions in the context that we shall be discussing them. Now, for the example, imagine that you are a tin can.

As a tin can you are well respected for your utilitarian purposes and capabilities of performance. True, you are taken mostly for granted, but you are needed and useful. On that day when you are opened and your content of beans is emptied into a saucepan, your existence takes a drastic new course. You are no longer respected. Your utilitarian purpose and ability to provide the services for which you were created, exist no more. You see that being taken for granted was far better than being unnecessary and unsightly in peoples' eyes. This condition is apparent to you now because you are a piece of solid waste. You have been *generated.*

Any number of generators could be your parent: a housewife in a home or apartment, a machine in a government cafeteria, or a family on a picnic in the park. But understand one thing quite clearly—you are unwanted.

In haste to get along with the beans, your generator immediately discards you to a temporary residence such as a paper bag, a plastic bucket or polyfilm bag, or the grassy knoll under the oaks beside the barbeque pit in the park. If your former protector was of a fastidious nature, the residue of beans may be washed down the sink disposal before you enter the next phase of existence. This temporary phase is probably short term if you are fortunate. You are *stored.*

Because you reside in storage next to other items of waste such as paper, plastics, glass bottles, rags, a broken plate, and leftover beans, you and your kind accumulate rapidly. Continued storage becomes unpleasant for your generator and you must be spurred along your way to someplace—preferably far away. So, you are *handled.* Maybe you are stored and handled in repeated steps. For example, you may be transported to the curb in your storage container and stored again for the trash truck, then dumped into the truck to reside again in a storage mode. Possibly, you are handled by being dumped while stored in a paper or plastic bag, down a trash chute in the building, and then find yourself in storage waiting to be handled by the building maintenance man who places you in a large bin for pick-up and dumping into the trash truck. Maybe a good soul picks you up, if you are in the park, and deposits you for temporary storage in the large wire basket.

But, regardless of where or how you have been stored and handled, you are still a tin can. You are all in one piece the same as the moment that you were generated. Maybe you smell a little bit or have a rust spot here and there, or are losing your label—but still a tin can.

A big event awaits you now. It's a type of handling, but it is a special type, because you are removed from the site of your generation and are *transported.* Your generator usually thinks of this phase of your existence as "the collection function" or most probably as "garbage collection."

Your generator makes a big thing about your being collected because you are now away from the site thankfully, even though it costs money to get you removed; and your generator can enjoy relief brought by the out-of-sight out-of-mind syndrome, a nearly universal affliction of generators. The municipally owned or private scavenger trash truck and personnel, however, have a lot more in store for you. You are about to be *processed.*

Actually, you are lucky. You could have been processed long ago. If you resided in storage in the home of an affluent American, you could have been compacted in a kitchen compactor, and within moments after you were generated. Or you may have been compacted or incinerated in the basement of the apartment building. If you had been flattened by a member of that family in the park, you could consider yourself as having been processed. To be processed, you realize, you have to change either physical or chemical form or have your environment substantially altered, as for example, being plucked from the meld of solid waste and placed with only other tin cans. Perhaps you are changed both ways, being segregated and compacted like they do automobile bodies.

Well, being processed is what usually happens to solid waste these days. Most tin cans get compacted on the site or more commonly in a collection truck. Some cans get burned. A very few get sorted out and recycled into a useful, needed, and respected product.

Possibly, because many of the public and private sector people who performed the handling or transport functions or the processing functions upon you, really care, you may undergo a *pre-disposal process.* That recycle process is one of the most exciting pre-disposal concerns these days, a thing that all the people are talking about. Maybe you can participate in some kind of handling, storage, and process function as part of a railhaul project that gets you better prepared for disposal far away. That might involve you with sorting, shredding, or baling for the long trip and might be done at a transfer station. So, you see that being processed has its good and bad points, and no matter how it's done, you wind up as hardly the shell of your former self.

Your ultimate end, unless you are processed for recycling, is the dreaded *disposal* function. Most generators call this the end of the waste problem. If you are incinerated, your shell still exists, but your solid waste friends are consumed and gasified. Often, the solid waste is not incinerated first, but all goes to its end by the disposal method called *sanitary landfill.*

This disposal business is grisly. But, fortunately, disposal is not the subject here. We are concerned with those basic functions of generation, storage, handling, transportation, and processing that are implemented by anyone and occur on-site, in transit, and in pre-disposal.

The levity expressed in definition of the collection function is intentional. The point is made that this very serious subject is both neglected and avoided because of a lack of information, stigma, and embarrassment. Particularly, the subject of on-site generation is a private affair to most citizens.[a]

The definitions employed throughout this study regarding the collection function may be objectively stated. These definitions include three principal participants who are concerned with four types of events and who perform five basic functions in the solid waste management cycle. The schematic diagram expressing these elements of the collection function is shown in Figure 1-1. It presents a view of the primary interfaces between participants, events, and functions, but is not intended to depict the full realm of possible interactions or dependencies. This schematic summary of solid waste management interaction may be simply stated as a pictorial display of who does what to solid waste and where. This book primarily deals with *how* the *what* functions are or may be performed.

In summary, then, the three types of persons involved, the four categories of location, and the five basic functions performed may be separately defined in the following ways:

Participants

The three principal participants in the solid waste sequence are:

1. Waste generators—all people
2. Municipal entities—collectors, incinerator and landfill operations
3. Business sector entities—private collection services, private disposers, and equipment makers

---

[a]By a 4 to 3 decision, the California Supreme Court ruled that curbside garbage containers are within a householder's "expectation of privacy."

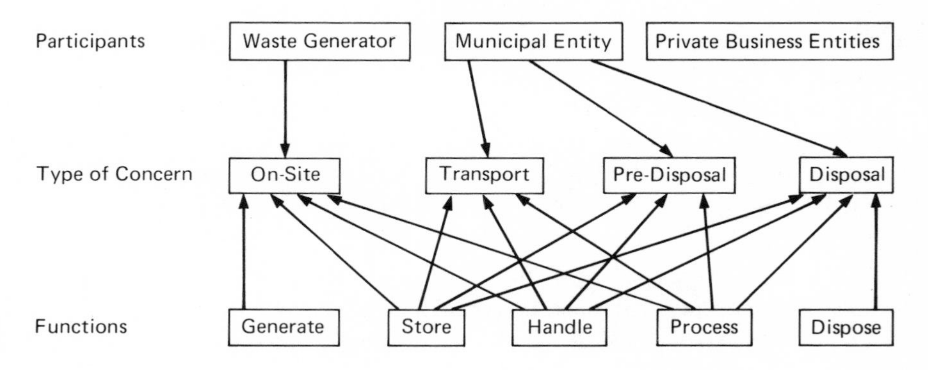

Figure 1-1. Solid Waste Interactions in Collection and Disposal.

Note that waste generation occurs at a site or location. The term "on-site" describes the location where waste originates. The following list of types of on-site generators covers most conditions of generation.

1. Domestic–single and multifamily dwellings
2. Multipurpose–the office/dwelling/restaurant building
3. Commercial–office or retail/wholesale business
4. Industrial–manufacturing of products occurs
5. Hospital–unique wastes are generated
6. Parks and public places–streets and litter
7. Agricultural–not a primary contributor to municipal load
8. Demolition–old and new structure materials
9. Bulk–automobile hulks or nonstandard domestic
10. Other

Events

The four major event categories in the solid waste sequence are:

1. On-site concerns
2. Transport concerns
3. Pre-disposal concerns
4. Disposal concerns

These major event categories form a sequence. The interrelation of the categories is displayed in Figure 1-2.

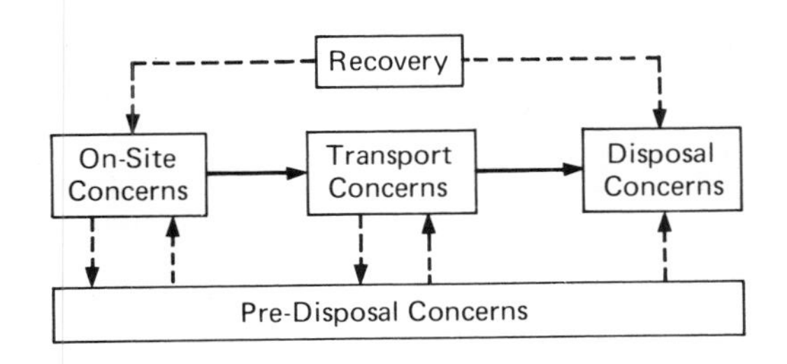

Figure 1–2. Interrelation of Major Events in Solid Waste Sequence.

Note that the solid waste flow is indeed sequential as shown by the solid line—from generation or point of origin the waste is transported to its destination for disposal. Historically, there has been little emphasis upon the events, and supporting activities to those events, that might beneficially effect a cyclic pattern instead of a sequential series of events, for the product moving along the lines in the block diagram—the solid waste itself.

Consequently, the numerous concerns for resource recovery, devising methods of greater economy or efficiency, and utilizing available technology to formulate better policies and procedures for *management* of solid waste, afford the creation of a new category of events. These are termed "pre-disposal concerns." Inclusion of these pre-disposal concerns in the waste movement chain has an impact on the participants and functions as suggested by the dotted lines in the block diagram; note that a cycle may be established from the former sequence.

These pre-disposal concerns establish a framework for better management of the solid waste flow, because cyclic functions are more sensitive to control operations than are the primarily rate oriented sequential flows.[b]

Functions

The five basic functions in the solid waste cycle are:

1. Generate
2. Store
3. Transport

[b]Numerous practical and theoretical discussions of management control theory are available. For example, see W. Hartman, H. Matthes, and A. Proeme, *Management Information Systems Handbook,* McGraw-Hill, 1968.

4. Process
5. Dispose

These functions are nearly self explanatory. It is notable that generation is an instantaneous occurrence or point function. It may be thought of as virtual creation at a discrete point in time. The other functions consume real and elapsing time for occurrence and are therefore time related functions.

The storage function is particularly important because space or volume must be available and utilized for its occurrence.

The handling function is a labor, capital, and energy intensive function. The waste itself is not physically or chemically altered.

The processing function is also intensively consumptive of resources of labor, money, energy, and time. Processing does alter the physical or chemical composition of the waste stream.

The disposal function is not an integral concern of this book. Neither is it a final solution to any aspect of the total solid waste management problem.

2 Solid Waste Characteristics

To fully understand the refuse collection functions, it is necessary to know a great deal about the sites where the refuse is generated and the viewpoints from which the refuse character may be examined.

The primary source of refuse generation with regard to tonnage each year, is from domestic sites. A recent study of solid waste problems in New York City [1] showed that 51.7 percent of municipal refuse loads originate in residential sites, 38 percent from commercial sources, and the remaining 10.3 percent from industrial generators.[a] Throughout the country, the nature of residential sites is shifting to the extent that the majority of Americans now live in multifamily dwellings.[b] This shift in residence type is continuing and effects change in the character and characteristics of generated refuse. This is readily apparent when viewing the population densities and demographic character of communities of people. For example, the single family residences average 9.0 people per acre, townhouses 47.7, garden apartments 66.5, medium-rise buildings 148.5, and high-rise structures 223.5 people per acre. Within these dwelling units, the average number of persons per unit ranges from 3.6 in single family residences to 3.3 persons in the medium and high-rise buildings.[2]

Most authorities agree unequivocally that population character, population density, and personal income levels are key indicators of per capita refuse generation.[3] Methods have been developed to establish baseline characteristics for municipal generation loads and to project the related factors that influence generation rate. These principal influencing elements of domestic generation are population, employment, income, and productivity. These determination and forecast methods are part of the study of demography and consider age, sex, migration rates, birth and death rates, and ethnic backgrounds. Features of non-manufacturing and manufacturing employment, construction employment,

[a] These data were achieved from 1967 surveys and statistical records of the New York City Department of Sanitation. Projections of these base calculations indicate that the source meld of the New York City municipal refuse load in 1975 will be 53.4 percent residential, 37.5 percent commercial, and 9.3 percent industrial. By year 2000, the meld is predicted to be 57.6 percent residential, 36.7 percent commercial, and 5.7 percent industrial.

[b] Based upon predictions by United States Census of Housing, 1970, and statistics accumulated by housing industry associations such as National Association of Home Builders.

output per man-hour productivity,[c] packaging materials consumption, and gross national product are included in a developed "model."

The term "model" means a mathematical model or sets of mathematical equations which use one or more variables to predict one or more unknown variables. A commonly used mathematical model is the type employing linear regression analysis. These techniques may allow substantiation of the method and data used in predicting future waste loads by checking the model accuracy with historical information. For statistical assurances that predictions are accurate, levels of confidence and correlation factors are developed which indicate the degree of trust that may be placed on such hypothetical determinations. Similar but less extensive and refined techniques are utilized in predicting both commercial and industrial refuse loads.[d]

In more tangible terms the quantities of municipal waste loads and the rates of generation of refuse may be examined in the literature.[4] It is usual to find these discussions jointly considering both quantity and generation rates, together with the quality or composition characteristics of wastes.[5] Numerous statistical sources,[6] technically oriented analyses,[7] and studies of unique wastes [8] are available; but, in lay terms, the on-site character of solid waste, regarding quantity and quality, may best be understood by examination of the refuse generator's environment and the type of site where wastes are created.

Nature of Domestic Refuse

Domestic refuse is a composition of organic food wastes; paper and paper products; wood; plastics, leather, and rubber materials; rags and textile products; glass; metallics; inert stone, clay, and earthen products; and yard and other bulky wastes. The usually accepted method for specifying composition is to express it as either percentage by weight or percentage by volume.

A study of the composition of domestic refuse was made in 1970 in low-income, public, high-rise housing in New Haven Connecticut, by a National Academy of Sciences study team that hand-sorted 11,000 pounds of refuse.[e] The results of that study are shown below.

---

[c] Productivity is a dimensionless ratio or index used by the U.S. Bureau of Labor Statistics to compare current productivity with base years' productivity, usually the 1957 to 1959 time period.

[d] Correlation coefficients of 0.97 for residential and 0.786 for commercial/institutional refuse generation were demonstrated in New York City studies.

[e] The source for these data was "Collection, Reduction, and Disposal of Solid Waste in High Rise Multi-Family Dwellings," An interim report by the National Academy of Sciences, Prepared by the Building Research Advisory Board Committee on Solid Waste, Washington, D.C., National Academy of Sciences, 1970, p. 21.

Category of Refuse	*Percent by Weight*	*Percent by Volume*
Paper and Paper Products	32.98	62.61
Wood and Wood Products	0.38	0.15
Plastic, Leather, and Rubber Products	6.84	9.06
Rags and Textile Products	6.36	5.10
Glass	16.06	5.31
Metallics	10.74	9.12
Stones, Sand, and other inerts	0.26	0.07
Garbage (organics)	26.38	8.58

However, no generally accepted data for composition of domestic refuse is available and various authoritative sources [9] and knowledgeable expert studies [10] derive analogous but different discrete data. It is explanatory, then, to identify reasons for the apparent discrepancies or variations and to avoid specific numerical definitions beyond examples.

First, the sampling and analytic techniques for examination of domestic refuse vary and are not standardized.[11, 12] Many governmental, professional, and industrial groups have defined categories of wastes and measured composition, but most sources define both categories and generation data in terms of the total municipal load and not in the more finite terms of its domestically contributed content. According to one report presented in 1968,[f] the average amount of waste collected per person in urban areas was 5.72 pounds per day. Rural area collection per person was only 3.93 pounds per day. It must be remembered that this included all sources of collection, and not just the home. The American Public Works Association[g] and Incinerator Institute of America[h] classify refuse quite differently and neither categorization segregates the domestic constituent from the defined municipal load. The former source defines all types of solid refuse—domestic, commercial, and industrial—in accordance with Table 2-1. The latter source defines all types of solid waste into the six categories shown in Table 2-2. It is, therefore, apparent that both definitions of waste composition and the methods for measuring, the physical and chemical content of refuse, vary widely. It is true, as well, that vague distinctions are made between

---

[f]R.J. Black, A.J. Muhic, A.J. Klee, H.L. Hickman, Jr., and R.D. Vaughan, "The National Solid Wastes, and Interim Report", Presented at the 1968 Annual Meeting of the Institute for Solid Wastes of the American Public Works Association, Miami Beach, Florida, 24 Oct. 1968.

[g]"Refuse Collection Practice", Third Edition, American Public Works Association, 1966, p. 15.

[h]Incinerator Standards, Incinerator Institute of America, Nov. 1968.

Table 2–1
Refuse Categories, Content, and Sources by American Public Works Association

	Category	*Content*		*Source*
Refuse (Solid Wastes)	Garbage	Food Wastes Market Refuse		From: households, institutions, and commercial concerns
	Rubbish	Combustible (primarily organic)	Paper, cardboard, cartons Wood, boxes, excelsior Plastics Rags, cloth, bedding Leather, rubber Grass, leaves, yard trimmings	
		Noncombustible (primarily inorganic)	Metals, tin cans, metal foils Dirt Stones, bricks, ceramics crockery Glass bottles Other mineral refuse	
	Ashes	Residue from fires		
	Bulky Wastes	Large auto parts, tires Appliances Furniture Trees, stumps		
	Street Refuse	Street sweepings, dirt Leaves		From: streets, sidewalks, alleys, vacant lots, etc.
	Dead Animals			
	Abandoned Vehicles			

Construction and Demolition Wastes	Lumber, roofing Rubble	
Industrial Refuse	Solid wastes resulting from industrial processes and manufacturing operations	From: factories, power plants
Special Wastes	Hazardous wastes: pathological wastes, explosives, radioactive materials Security wastes, negotiable papers, etc.	From: Households, hospitals institutions, stores, industry
Animal and Agricultural Wastes	Manures, crop residues	From: Farms, feed lots
Sewage Treatment Residues	Coarse screenings	Sewage treatment plants septic tanks

Table 2–2
Refuse Categories, Content, and Sources from Incinerator Institute of America Standards

Type	*Description*
0	Trash, a mixture of highly combustible waste such as paper, cardboard, wood boxes, and combustible floor sweepings, from commercial and industrial activities. The mixture contains up to 10 percent by weight of plastic bags, coated paper, laminated paper, treated corrugated cardboard, oily rags, and plastic or rubber scraps.
1	Rubbish, a mixture of combustible waste such as paper, cardboard, cartons, wood scrap, foliage and combustible floor sweepings, from domestic, commercial, and industrial activities. The mixture contains up to 20 percent by weight of restaurant or cafeteria waste, but contains little or no treated papers, plastic or rubber wastes.
2	Refuse, consisting of an approximately even mixture of rubbish and garbage by weight. This type of waste is common to apartment and residential occupancy, consisting of up to 50 percent moisture.
3	Garbage, consisting of animal and vegetable wastes from restaurants, cafeterias, hotels, hospitals, markets, and like installations.
4	Human and animal remains, consisting of carcasses, organs and solid organic wastes from hospitals, laboratories, abattoirs, animal pounds, and similar sources, consisting of up to 85 percent moisture.
5	By-product waste, gaseous, liquid or semiliquid, such as tar, paints, solvents, sludge, fumes, etc., from industrial operations.
6	Solid by-product waste, such as rubber, plastics, wood waste, etc., from industrial operations.

Source: *Incinerator Standards,* New York, Incinerator Institute of America, Inc., 1968, p. 5a.

municipal wastes characteristics and the nature of constituent elements attained from domestic, commercial, and industrial sources.

Secondly, domestic waste generation varies seasonally [13] as is shown in Figure 2–1. It is readily understandable that yard waste contribution to the domestic load increases during summer months. It is also verified by monitoring of on-site sources that waste production varies in composition with season—percentages of paper rise to extraordinary levels during Christmas holidays. Consumer buying practices vary by day of the week and this reflection on waste generation rate is distinct and substantial. [14] Therefore, spot determinations of generation rates and composition of domestic refuse must be avoided to preclude erroneous interpretations drawn from such data.

A third caution regards the geographic source of data relevant to waste character. The average per capita generation of wastes in the West Bengali village of Deshapera is 0.7 pound per day, well below the commonly quoted

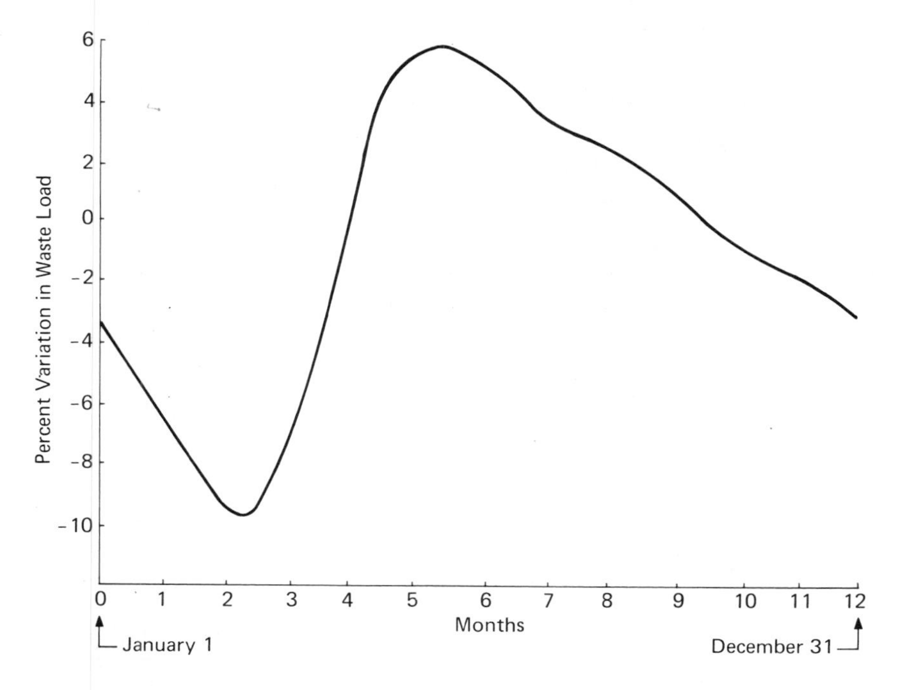

Figure 2-1. Seasonal Variation in Waste Load. Source: Envirogenics Co., "Systems Evaluation of Refuse as a Low Sulfur Fuel," A final report to the Environmental Protection Agency, Vol. 1, Report No. F-1295, November, 1971, p. II-16.

figures of 1.5 to 6.0 pounds per capita in the United States. The refuse is less organic than those wastes generated in India's urban areas,[i] which is understandable, due to the very poor economic conditions of the rural population.[15] In the United States, the geographic variations that affect waste character are not well defined; studies by University of Florida indicate that per capita generation is 4 percent higher in Gainsville, Florida than in Minneapolis, Minnesota,[16] and that composition does vary from city to city within the nation.

A fourth and principal reason for the cited differences in domestic waste character regards the economic level of the community wherein measurements are made. In those areas where affluence allows purchase of products with a high

[i]The uncompacted density of these rural measurements indicates a 23.3 pound per cubic foot average.

degree of packaging, the composition of wastes may be sharply altered. It may be safely stated that the higher the disposable income level, the higher is the paper content of the refuse, and the lower the income level, the lower the food waste content of the refuse.[17] After World War II, the refuse per capita generation in Europe closely followed the economic recovery on the Continent. Annual per capita generation has risen from 475 pounds in 1953 to 725 pounds in 1964, while densities have decreased in the period from 485 pounds to 390 pounds per cubic yard, indicative of increased paper and packaging content of wastes.[18] This is equivalent to 14.4 pounds per cubic foot in density and substantially more dense than American refuse.[j]

The connection between waste character and personal income level has been discussed at length [19] and has been historically proven on a worldwide basis. It is anticipated that by the year 2000, the domestically created and total municipal waste load will increase by 270 percent [20] per capita.

Nature of Commercial and Industrial Refuse

Although domestically created refuse is the majority contributor to the municipal waste load, the numerous other on-site generators deserve equivalent if not greater attention and study. The problems associated with nondomestically generated wastes are more complex due to factors such as the uniqueness of the refuse—such as in hospitals or with dangerous industrial materials—and the dispersion of the wastes, as with roadside litter, or with the shear bulk of the problem, as arises with abandoned vehicles.

A new type of urban structure presents unique waste generation characteristics emanating from the multipurpose building. The quantity and quality of refuse generated in the office/dwelling/restaurant structure presents the need for specialized viewing, but little directed study has focused on this new and growing source. The more traditional approach to viewing waste character does not examine this combination source and instead views the commercial and industrial elements separately.

Commercial refuse generation is primarily an urban matter.[k] More recent

---

[j]The previously cited National Academy of Sciences study examined domestic refuse densities on-site. An average value of 5.6 pounds per cubic foot has been measured with weekly averages ranging from 4.7 to 6.7 pounds per cubic foot. Daily averages by hour in these tests indicate density values covering the spectrum from 4.0 to 7.6 pounds per cubic foot.

[k]The U.S. Public Health Service review of solid waste composition from various sources in urban and rural locations conducted in 1968 ascribed per capita per day contribution from commercial sources as 0.46 pound in the urban and 0.11 pound in the rural sector.

study of the commercial contribution of waste to the municipal load has provided methods for estimating rate generation for categories of business and commerce.[21] This study indicates that, for estimative purposes, solid wastes generated from commercial establishments can be mathematically related to characteristics of the establishments under consideration. The generation of commercial solid waste was found to be most closely related to the number of employees, hours open, and type of establishment involved. The variable system employed indicated that for restaurants and clothing and hardware stores, the relationship between waste generation, number of employees, and hours open was similar. For the remainder of the common commercial establishments, drug and grocery stores, however, the type of establishment involved should also be considered.

The composition of commercial waste is primarily food and food preparation wastes and the paper and box wastes from the retail business activities. The specialized institutional wastes, however, within the commercial category have a unique character. Hospital and medical center wastes pose a new and rapidly growing class of refuse containing many disposable metal and plastic products, in addition to bacterial and viral constituents to organic waste.[22] The total waste load from hospitals may be related to the number of beds available; projections of these data by various hospital associations and medical groups indicate a widening gap in the nonlinear relationship as shown in Figure 2-2. In 1968 for 6134 beds in a sampling of 17 hospitals, 22.2 tons per day of refuse was generated. By 1983 these facilities are projected to grow to 9400 beds and will relate to 56.0 tons per day of generated refuse.[1] The quantity and character of this category of institutional waste is rapidly changing and growing, to eliminate internal handling problems and the associated hazards in the spread of disease. A recent survey of twelve hospitals showed the composition to be 127 disposable items in nursing services, 29 by dietary services, 24 in surgical and obstetrical use, 26 in laboratories, and 13 in housekeeping departments.

Industrial wastes contribute the second largest single category of waste produced in the nation and the identification and handling of these sources of refuse is very complex.[23] Public Health Service statistics from 1968 attribute the distribution as 0.65 pound per capita per day from urban located sources and 0.37 pound from rurally located generators. These sources are many and quite diverse in contribution to refuse load. Recent studies conducted under contract with the Council On Environmental Quality, the Environmental Protection

---

[1] "Medics Face Giant Problem in Refuse Handling", *Solid Wastes Management/Refuse Removal Journal,* 14 (3): 46, March, 1971.

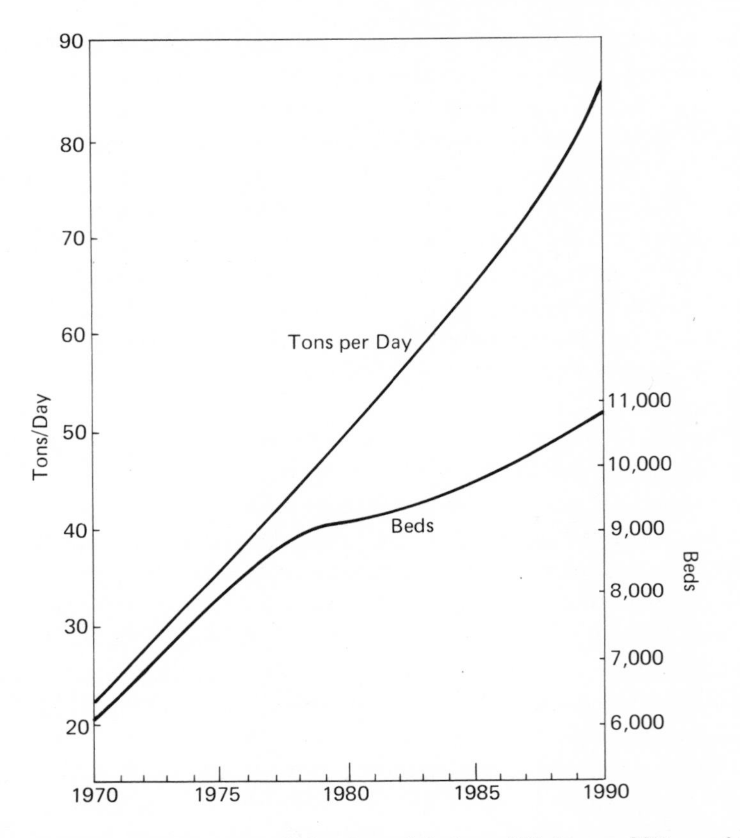

Figure 2-2. Estimate of Number of Beds and Volume of Disposables Generated for a 17 Hospital Sample. Source: "Medics Face Giant Problem in Refuse Handling," *Solid Wastes Management/Refuse Removal Journal*, 14 (3): 46, March, 1971.

Agency, Department of Interior, and Council of Economic Advisors[m] have only begun to measure and evaluate the character and impact of industrial waste generation through micro- and macroeconomic analyses. These studies, however, have not examined specific solid waste load character or contribution from the industrial sector, and little information is available other than to identify metals, wood, paper, glass, rubber, and textile constituents as predominant. Little organic food wastes are present. Industrial sources salvage on-site those materials

[m]See "The Economic Impact of Pollution Control", A Summary of recent studies, from Environmental Protection Agency, March 1972.

that are economically advantageous to reclaim.[24] One of the best available descriptions of the industrial waste load character [25] was prepared by the predecessor agency of EPA's Office of Solid Waste Management, the Bureau of Solid Waste Management of the Department of Health, Education and Welfare.

Nature of Refuse in Public Places

A very substantial element of the total waste load comes from litter [26] and bulky wastes—including automobiles—generated in parks, road rights-of-way, and city streets. Littering is a practice of young people primarily and decreases with increasing age. Among adults between the ages of 21 and 35, three times more litter is generated than from those persons over 50, and twice as much as those in the 35 to 49 age group.[n] The content of litter is primarily paper and cans.[o] On a daily park maintenance tour through Washington, D.C.'s Rock Creek Park, 117 cubic feet of litter are collected from roadside and public grounds amounting to 857 pounds.[27] It appears that total litter volume is solely and positively correlated with average daily traffic flow[p] and not related to the type of road or public place. No one knows what the per capita generation of litter really is, but we can estimate that the cost for its removal is enormous.

With scrap steel prices low and transportation costs high, the abandoned automobile problem is mounting in most major cities.[28] Salvage of old cars will seldom pay for their removal and disassembly costs, as well as produce a profit. Some estimates are that over a million cars are simply abandoned by their owners each year.[q] In 1969, New York City removed nearly 60,000 junk vehicles.

The composition and density of solid waste products are important factors in the collection and transportation functions. Whereas waste composition is important to recycling, incineration, and landfill efforts because the percentages of various materials must be known, the collector is concerned with composition for other reasons. If a major portion of the collected trash is stacked newsprint

---

[n]Data acquired and documented in November 1968 by Public Opinion Survey, Inc., "Who litters—and Why".

[o]As reported by J.C. Markens, in "Paper and Beer Cans Win", *American City,* January 1970, a good indication of litter composition is 57 percent paper, 36 percent cans, 7 percent other materials—primarily glass.

[p]"National Study of the Composition of Roadside Litter", by Research Triangle Institute, Research Triangle Park, North Carolina, September 1969.

[q]Abandoned Automobiles—The Problem, Solid Waste Management/Refuse Removal Journal, April 1971, p. 18.

Table 2-3
Quantity per Materials Balance Analysis on an as Discarded Basis (X 1,000 tons)

Material	*Refuse*	*Percent of Total*
Paper	38,200	33.3
Glass	10,800	9.4
Ferrous	8,900	7.7
Aluminum	630	0.55
Tin	66	0.06
Copper	355	0.3
Lead	30	0.03
Textiles	2,900	2.5
Rubber	1,600	1.4
Plastics	2,200	1.9
Food, Animal, Plant, and Other Wastes	49,319	42.9
Total	115,000	100.0

and flattened cardboard containers, there will be little advantage in attempting to compact the material in the collection truck. Conversely, loose mixed trash may be compacted to decrease the total number of trips required to the point of processing or disposal. The density of refuse is important to the collector because the weight per cubic foot determines the methods of movement from the point of generation.[r]

As we understand more about how the refuse collection functions are carried out, it will become all the more apparent why we must know and be able to predict the nature and character of solid wastes [29] that are part of the municipal waste load.

The National Center for Resource Recovery (NCRR) has thoroughly analyzed this question and Table 2-3 gives their projections of the composition of municipal solid waste on a nationwide basis. For a more complete analysis of their conclusions, see Reference 30.

[r]New York City Public Law 14 imposes limits on the generators of waste materials by precluding use of garbage disposals in residences, hence more organic wastes in refuse, and limits the density to 700 pounds per cubic yard allowable for city collection.

3 On-Site Concerns

Within the framework of conducting the storage, handling, and processing functions at the site of waste generation, numerous conditions require examination and evaluation. As defined in the preceding chapter discussing the character of refuse, we see the specialized points of view needing attention regarding the systems and equipments available, facilities needs, personnel and safety practices, the responsibilities for completion of tasks, and the costs involved with conducting on-site refuse management. Indeed, without proper storage and handling of refuse at the site of generation, we would soon be deep in our own wastes. [1] Times have changed since livestock was used as garbage disposal units. [2]

Historically, lower population densities allowed disposal of waste products near the source. [3] The availability of vacant areas amid the low concentration of population, coupled with smaller amounts of waste products, allowed the individual producer to throw garbage onto the back part of a lot or into some nearby ravine, field, trench, or other unused area.

The raising of domestic animals was not uncommon, even in fairly well populated areas, and these animals often provided processing of garbage with any food value. [4] Fireplaces, coal-burning stoves, and furnaces usually provided the incinerator required for disposal of paper and other combustibles. In fact, the shortage of highly combustible refuse often forced the homeowner to purchase kindling, cobs, or other materials to start fires in furnaces, stoves, and fireplaces. Even the archaic kitchen range provided convenient disposal for sacks and other wrapping materials. The ash and cinders from such burning provided fill for driveways, alleys, or were thrown on streets during the winters.

Larger noncombustibles usually presented little disposal problem. Scavengers with horse-drawn or motorized vehicles made occasional forays up the alleys seeking metal objects of scrap value or other junk that could be rehabilitated. These same alleys provided a disposal area for some less objectionable waste. When sizeable portions of the populace could not afford new washers and refrigerators, it was a simple matter to give such items to the family who would repair them and make do with castoff appliances. Before World War II, the economic conditions also made the hand collection and sorting of scrap metals, rags, and glass a profitable venture for certain persons. With the entry of the U.S. into the war, lack of new consumer goods assisted in keeping discards to a minimum. Scrap metal, paper, rubber and grease collection drives also provided a means of discarding much refuse. While this method of using or recycling solid waste may not have been economically practical, it served a useful purpose at that time. Food rationing also played a small part in controlling the amount of organic waste.

It was not until the recovery years after World War II that the solid waste handling problems became really pronounced. As saturation of the potential auto market appeared imminent, and the buying power of the individual was supplemented by easy credit, the buy and discard philosophy increased. Increasing sales were accompanied by increased use of packaging. The prepackaging of groceries, with the rapid growth of supermarkets and convenience foods, added to the problems. Increasing labor rates in the service field promoted a buy rather than repair philosophy. The increase in population, coupled with apartment and subdivision construction, raised the persons per square mile ratio in the metropolitan areas. The elimination of alleyways in new development planning increased the visibility of solid waste. Homeowner associations assisted health agencies and city councils in forcing ordinances controlling the frequency of collection and limiting the quantity of on-site wastes. However, the affluence of today's American society has initiated new on-site waste problems.

The first pressing on-site refuse problem regards storage. On the domestic scene, the problems of storage capacity, odor, and vector control make the storage task of initial and prime importance in residences.[5] Without proper care in storage of home refuse, insects[a] and vermin[b] problems become intolerable.[8]

On-Site Storage

Various on-site containers are used on residences to store solid waste. However, they may generally be divided into two categories—(1) the metal or plastic, reusable containers that are returned to the owner, and (2) the plastic or paper container, usually a bag, that is collected with its contents for disposal. Bags are fairly new innovations in the area of solid waste collection;[9] returnable containers have been with us for some time. Both paper and plastic refuse

[a]Results of Southern California studies, as reported in Reference showed that household refuse storage containers are a major source of green blowflies even though the area has moderately cool weather during most of the summer.

[b]As described in Reference 7, a continuing program of inspection, code enforcement, general environmental sanitation, education, and rat-proofing must be developed in and for American cities regarding problems of on-site waste storage.

storage bags are increasing in use on a national basis.[10] This condition arises because sealed bags promote better health and cleanliness at the site and allow hygienic collection and disposal practices, efficient use of labor, noise reduction, and improved appearance.[11]

Paper bags are being adopted by various cities. In some cases, the bags are furnished at no charge as a part of the municipal trash collection service.[12] The costs of such bags are covered by the funds allocated for the service and ultimately paid for by taxes or other assessed fees. In other cases, the householder pays for the bags used.[13] Most residents of a community are quite happy with the paper bag storage concept due to the small expense involved.[c] In fact, a country-wide survey of 1856 homes, as reported by the American Public Works Association, indicated that 88 percent of the homeowners favor the use of disposable paper or plastic sacks over the conventional refuse storage container systems used on-site.[14]

The City Manager of Weatherford, Texas, has described the success of that city's paper bag program.[d] Prior to the introduction of paper bags, Weatherford provided twice a week carry out trash service for $1.50 per month for each resident.

The city then decided to perform a 217-home test with paper bags. The paper bag storage apparatus consisted of metal frame holders with plastic lids for securing the bags behind the houses. The filled bags were then placed at curbside for collection.[e] After the 90 day trial period, the residents of the test route were queried as to the desirability of continuing such a service. Positive acceptance of the system was received, and this has since been expanded to 4,000 homes.

Bags have other advantages over metal trash cans. They provide a quiet means of transferring trash from the residence to the collection truck. Bags are usually offered in a 30 gallon size and are constructed of paper sufficiently rugged to withstand the rigors of their handling as well as their contents.[15] Chemical impregnation of the paper protects it from water and grease. Repellents are also being added to the paper to discourage hungry animals. While metal trash cans may be reused—their life is definitely limited by rough treatment. The life cycle of the disposable bag, either paper or plastic, is limited to one trip.

The plastic bag for home refuse storage serves the same function at approxi-

---

[c] More than one-fourth of the 4000 household residents of Camden, Arkansas, elected to pay $0.65 extra per month for paper bags for on-site refuse storage and collection.

[d] K.E. Smith, "The People Chose Sacks over Cans", *The American City*, June 1970, p. 82–83.

[e] A reduction of 66 percent in collection labor on the trial collection route was achieved as a bonus to the paper bag program results.

mately the same cost as the paper bag. It is usually 0.0015 to 0.004 inch thick polyethylene with the thinner bag material costing about one-half as much as paper bags, and the thicker bags approximating the cost of paper bag containers. New innovations in plastic bag design include a butyl plastic bag, and another approach incorporates a butyl with chemical additive, to cause degradation and decomposition of the bag several months after use. A self-standing polyethylene refuse bag is now available for use in the home as a kitchen container, storage container, and disposable refuse bag for collection [16] and is specified for use in several American and Canadian cities. However, many cities do not allow use of plastic and/or paper bags and applicable codes of cities are changing to permit use of disposable bags.[17] Ordinance changes are being made[f] and suburban as well as city dwellers are greatly pleased [19] with cleanliness and economies.[20]

Americans are not the only beneficiaries of the bag storage system.[21] Some European cities make the paper sack storage and collection mandatory—in keeping with the new town concepts of modern, British city planning.[22] In Florence, Italy, nearly 100,000 households have been utilizing paper sack storage on-site for five years.[23]

Adoption of trash bags as a means of containing solid waste has met with favorable acceptance. In every case, the advantages of bags over cans can be noted from the viewpoint of both householders and refuse collectors. A partial list of these advantages includes:

1. Pickup personnel are not directly exposed to the waste material. This positive aspect of bags eliminates much of the filth that was previously encountered by the collection personnel.
2. Hygienic problems, created by organic materials that are normally encountered in the residue left in the cans, are eliminated.
3. There is a total reduction in collection time. The container does not have to be returned to curbside or the residence. This elimination of an extra trip at each stop allows the work crews to make more stops in a given time. Reduction of time by 50 to 60 percent is commonly noted.
4. The cost of the bags is partially overcome by the elimination of reusable containers. Even heavy metal cans must be periodically replaced.
5. Blowing trash is reduced. When the amount of solid waste produced exceeds

[f] As described in Reference 18, the National Science Foundation sponsored study and final report has evaluated the advantages and disadvantages of disposable storage containers. The Disposable Refuse Container Project conducted by New York City has proven successful and legal changes were adopted.

the capacity of the available cans, the householder tends to deposit trash on the ground near the cans, or stacks material on top of the cans.

6. Space requirements are reduced with bags. A metal or plastic container consumes the same amount of space, whether it is full or empty. An empty bag requires very little space. The flexibility of the bags allows them to be stacked or packed next to each other.
7. Bags are also useful for other types of refuse. Grass clippings, leaves, and other yard vegetation may be placed in bags. This procedure eliminates burning and also prevents wind scattering from raked piles of refuse.
8. Uncovered trucks could be used to haul refuse in some cases. The throw-away container prevents dispersal of trash during transit.
9. Social acceptability of bags is greater. They are less of an eyesore on collection days. They do not create the noise of cans during handling by pick-up crews.

Often the location of storage containers is specified by law. If it is not, the refuse collectors will usually assist the homeowner and indicate preferred storage containers, location, and availability to meet specific storage requirements. When requested, their services can be quite helpful.[g]

Modified trash cans[h] are being marketed for residential use that differ slightly in utility from standard devices, but have some of the same disadvantages. [24] Recessed trash and garbage cans have been used to eliminate the unsightly appearance of aboveground containers. These recessed cans are usually set in a concrete chamber or vault, and have a heavy lid that prevents trash blowing or animal scavenging. An obvious disadvantage is that the collector has to lift the full container out of the hole. This is cumbersome and physically taxing if the contents are heavy and difficult to clean.

Trash cans with wheels and a handle that allow the homeowner or collector to roll rather than slide the container have also been marketed. Containers have been manufactured from metal and plastic, square shapes have been used, eye appealing colors are being tried, and other innovations have emerged. However, no revolutionary improvements have been made in this method of on-site storage.

A possible approach to decreasing costs of solid waste storage handling at the source is found in containers shared by two or more residences. In such an

---

[g] "Cans: Where to Place Them", *Solid Wastes Management/Refuse Removal Journal,* 13(3):26–40, March 1970.

[h] Los Angeles now requires cans to be from 20 to 45 gallon capacity, circular and tapered in shape, constructed of metal or plastic and possessing handles.

operation, large enclosed containers are conveniently placed near the sharing homes.[i] Each resident removes trash from his premises, places it in the container, and a single operator truck with a mechanical lift mechanism periodically collects the refuse.

While this approach is helpful in limiting labor involved in collection, it has certain disadvantages. The container, being sufficiently large to hold the contents of ten trash cans, has high visibility. Ideally, a residential area served by alleys would be adaptable to such an approach. Some degree of cooperation is needed among residents being served by any one container to limit scattering or overloading. Landscaping via shrubbery around the container would be an asset. Another problem arises in placement of the container. The advantage gained in decreased pickup time may be far outweighed by the disadvantage of poor aesthetics and the difficulty of getting the household refuse to the container.

Trash pits, bins, vaults or large fixed metal, masonry, or wood containers have been discouraged as an on-site solid waste storing receptacle. These require the collector to shovel the contents into the collection truck. Not only are such pits dirty, but they also encourage odors, rats, fires, scattering, and expose the collector to the contents.

Due to high population density in multifamily residences, the total waste load of a dwelling is often quite large. Such buildings often require daily removal of stored quantities of refuse.

The large steel trash bins with wheels to facilitate moving are a popular item in many apartment operations. These bulk storage bins are commonly one-half to five cubic yards in capacity. Some collection services utilize trucks with lift-tilt mechanisms that preclude handling by the operating crews. Such loaders simplify the transfer of the trash from the on-site enclosure to the truck and markedly decrease the time for collection operations of refuse from storage.

An even more utilitarian modification of the described enclosure is a metal container that is really an enclosed truck body (less the wheels). This enclosure is slid or winched onto the truck bed and becomes the refuse vehicle to the final disposal site. These storage vans usually are twenty to forty-two cubic yards in capacity. Some systems simply leave the truck bed and body at the building site and this semitrailer is attached to the motorized cab for removal.

On-Site Refuse Handling

Occupants of low-rise multifamily dwelling units, such as garden apartments or the older two and three story apartment buildings with common stairs and

---

[i] M.D. Bogue, "Clean and Green Solid Waste System in Alabama", *Waste Age,* 1(5):4–6, 10–11, 36, September-October 1970.

interior or exterior corridor access to individual apartments, have a differing requirement in storage and handling of wastes. Although in some cases facilities provided for storage are comparable to those provided for single family dwelling, including rows of garbage cans in alleys or service areas, improvements such as common storage bins are gradually being adopted. In the case of buildings over two stories high, chutes may also provide vertical transport to a central storage room and/or bin. The conventional systems of medium and high-rise multifamily apartment buildings are generally limited to the trash chute for the vertical transport of refuse to a central storage room or processing unit. These vertical gravity chutes or tubes are commonly fabricated of aluminum, aluminized steel, or stainless steel and range upward in cost in the order named.

The chutes are commonly made in cylindrical form, as opposed to a square configuration, to provide greater strength with the use of less material. The cylinder is more readily cleaned and provides less probability of the accumulation of dirt and putrescible matter which might attract insects. Chutes are available in diameters from 12 inches to 36 inches, with 24 inches being usual. A disinfecting and sanitizing unit should be added at the top of each vertical riser in the trash chute system. These units are pressure-operated sprays which add disinfectants to the spray of water. The general cleanliness and odor-free condition of the chute is largely dependent upon the frequency of the use of the sanitizing unit.

Prices of basic standard chutes, excluding required building enclosures (chase walls), will vary from $140 to $270 per story, depending upon the materials of which they are made. Erection cost will be about $50 to $60 per story, including insulation requisite to meet building and fire code restraints.[j]

Actual on-site storage facilities in residential complexes and apartments are usually too small, inadequately designed for logistics of the collection function, and do not provide adequate supporting utilities and trained personnel to conduct the refuse management operation. The rooms where wastes are stored in apartments are characteristically small, possess inappropriately located utilities, have poor accessways for refuse removal, exhibit inadequate attention in architectural design for vertical transfer of wastes to storage, and are diminutive in waste management facility design to preclude loss of premium rate rental property.[k]

While barrels and sacks are used by some of the very small commercial establishments,[25] the quantities of waste produced by large businesses are

---

[j]These data are abstracted from Building Research Institute records and supported by information derived from the Environmental Protection Agency publication (SW-35c) entitled "Solid Waste Management in Residential Complexes."

[k]In the conduct of a national field survey for the National Academy of Sciences, Building Research Advisory Board, regarding building waste storage and handling facilities, these tentative conclusions were drawn. Publication of findings and conclusions is pending.

usually so great that bigger containers become a necessity. Such quantities of waste also require a more complex on-site waste storage and handling system than small business establishments. There is the possibility that waste from certain sources can present hazards not normally encountered in domestic refuse. Sharp metal objects from fabrication operation, broken glass, and inflammable material require special attention. The waste from some operations may have scrap value and may even encourage theft.

While institutional waste loads are special [26] and require specialized storage and handling precautions, usually the waste load is handled in a similar manner to large building residences and commercial establishments. Wastes are containerized in cans or bags when volume is not excessive,[27] but the general conditions of storage and handling require dumpster type containers. Hospitals, particularly, face unique storage problems [28] and must safely store, handle, and release for disposal the pathogenic and specialized wastes found only in this institutional environment,[29] currently estimated to exceed 7.0 pounds per patient per day.

Industrial waste containers are usually similar to those used in larger business buildings or multifamily domestic structures. Unless special containerization is required to hold caustic, inflammable, or otherwise dangerous waste, the concern becomes one of handling the resulting volume and density.[30] In numerous plants, the solid waste removal program has no contact with municipal systems, but depends on contractors or a company operated system. If a rail siding serves the plant, the waste may be loaded into cars for storage and removal. A plant may utilize an incinerator for combustible scrap and in this case the problem becomes one of storage and handling of the ash.

Storage and Handling of Refuse in Public Places

Storage and handling of refuse in public places is quite a difficult problem. Not only is the volume substantial but also the distribution of the waste load over vast areas contributes to the magnitude of the problem. Collection of litter is very expensive.[31] In the urban environment the cost of litter baskets and their collection may be defrayed by allowing private companies to place advertising on the containers.[1] Other nations of the world face similar problems;[33]

---

[1]New York City faces a $2 to $3 million investment in litter containers for street use as described in Reference 32, and paid advertising is utilized to recover costs.

however, Soviet streets are among the cleanest due to public discipline[m] and prohibition of litter.[34] Collecting litter is a labor intensive function at sites where it occurs; it usually falls to the public entities to conduct the function. A typical mile of U.S. highway acquires 1,304 items of litter monthly [35] and the cost of the collection of these items to storage is frequently as much as the collected item was initially worth.[36] The Michigan State Highway Commission expends $0.30 to clear away a discarded beer can—its approximate price including content at original purchase. Public events including crowds of spectators contribute great quantities of refuse and litter to the municipal load.[n] These items are usually collected by hand and transported in small containers or bags to a bin for truck collection. In larger areas such as public beaches, motorized trains of wheeled carts[o] are utilized in the public sanitation function.[38]

However, of all public refuse problems, the abandoned automobile presents the greatest on-site headache. Vehicles are unsightly, dangerous to traffic, and a source of vermin. Handling of abandoned vehicles to the point of reclamation or disposal has posed such difficulty that in some urban areas, a regional approach is being taken to address the common problem.[39]

Preparing for Refuse Removal

In viewing the on-site storage and handling of refuse, the carts, dollies, cans, barrels, bins, paper bags, plastic bags, and chutes used, are the traditional equipment and approach to waste management. With slight improvements and variations these methods have been utilized for decades. Some more substantial improvements were needed and have been made in the important area of moving and handling of refuse to the collection vehicles.[40] Most collection operators agree that architects and planners give little thought to this aspect of on-site waste management. Large office, commercial, residential, and industrial structures are seldom designed to facilitate a rapid and economical transfer of wastes and most lack suitable storage areas.

A fairly standard and costly practice is to store the waste in containers that must be lifted and carried to the loading hopper on the truck, return the empty

---

[m] Soviet authorities post names, photographs, and addresses of littering offenders in public places.

[n] The Public Service Director of Daytona Beach Florida reported the collection of 770 cubic yards and 260 cubic yards of compacted litter and refuse after the Daytona 500 and Firecraker 400 autoraces, respectively. See Reference 37.

[o] Trains and carts or dollies used in public litter removal and storage vary in size from one-half to five cubic yards in capacity.

containers to the source, and pick up other full containers and repeat the operation. The containers are usually far from sanitary, may be overfilled and overloaded, tax the physical ability of the handler, and may not be easily emptied as material will stick to the sides or be wedged in the can. The shortest possible path to the truck may be hazardous, requiring the crew member to go up or down stairs, pass between other vehicles, and have partly obscured visibility while carrying the cans.

New Concepts for On-Site Storage and Handling

On-site handling and transporting technology is experiencing a rapid change today with regard to the new vacuum collection systems. One of the greatest problems of collecting solid waste from the on-site generator is transporting it from the home, apartment, office, hospital room, or other relatively intermittent and low volume source, to a central high volume point where it can be processed or transported to disposal. The ideal collection system would be one that consists of a receptacle at the point of waste generation and an immediate means of transfer of the waste to some remote point. The entire system should be closed, preventing odors, scattering of waste, and health hazards. As such, a system would eliminate intermediate handling, it would replace collection vehicles and their operating crews, and would eliminate the high visibility of unsightly trash.[p] Such systems have been discussed, proposed, researched, and developed. A few vacuum collection systems have been installed and have not proven to be problem free. They consist of large ducts, similar to sewage systems, that incorporate suction and air pressure to pull the waste to a large trash "depot," much as a vacuum cleaner pulls dirt into its collection bag.

Sweden was one of the first countries in which pneumatic solid waste collection was used on a limited basis.[q] One early system was installed in a hospital at Solleftea in 1961 and served as a proving ground for a similar system installed in 1967 in a 2,770 unit, 9,000 person housing complex in Sundeberg. In this latter system, solid waste is drawn for distances up to 1 mile to a bulk incinerator. The refuse travels at speeds up to 90 feet per second in pipes as large as 24 inches in diameter to handle the waste. The Swedish system AVAC (Automated Vacuum Collection) was developed by A.B. Centralsug and is

[p]The unique element in the pneumatic vacuum collection concept is its lateral transfer capability. A discussion of this system approach appears in "Pneumatic Tubes for Garbage Collection Get U.S. Trial Run", *Product Engineering,* 41:16, March 2, 1970.

[q]A.J. Marchant, "1984 in 1967", *Compost Science,* Spring-Summer 1967, p. 19.

1970 · 1967

handled in the United States by Envirogenics, an operating company of Aerojet-General Corporation.[r]

Eight Centralsug systems are presently operating and fifteen others are under contract in Europe and South America. Envirogenics has installed an AVAC system in Disney World in Florida for approximately $1 million. This installation is capable of handling 60,000 to 75,000 pounds of waste per day, uses 20 inch pipe, and the two mile network links 16 collection points with the central collection station. The hotel is about one mile from the collection station and trash is transported over the route in one minute (or at the rate of 60 mph).

Hospital AVAC systems are presently being installed at 6 institutions in the United States. Two residential installations will be made—one in a 20,000 person community on Welfare Island in New York's East River and the other in the HUD, Operation Breakthrough Project at Jersey City, New Jersey. Individual residences in Stockholm, Sweden, have been compelled for the past 30 years by ordinance to have a central chute system amenable to the AVAC concept [41] in all new construction. Similarly, the system finds application in Germany.[42] Studies in this country have suggested the feasibility of replacing collection trucks over the next 50 years with a completely automatic, $88.3 million AVAC piping system.[s]

In another study performed by the staff of Combustion Power Company, Inc., of Palo Alto, California, for the Bureau of Solid Waste Management, it was suggested that a pneumatic collection system be combined with a high temperature combustion incinerator. The energy achieved from burning waste would be used to drive the compressor energizing the pneumatic collection tubes. This proposed VaCol (Vacuum Collection) system should be able to collect 400 tons of solid waste produced daily by 100,000 to 150,000 residents in a matter of a few hours of operation.

The basic components of the VaCol system are the vacuum pumps, solid waste storage area, underground pipe network, collection valve, and the wall mounted deposit chutes. Combustion Power Company believes it is neither necessary nor economical to run the pump at maximum pressure continuously for efficient collection. Therefore, the system is operated intermittently and collection is accomplished by sequentially opening the valves at each residence during the operation periods. Thus, waste is temporarily stored in the feed pipe

---

[r]Such types of mechanization seem to be viable developing technology. A discussion of this subject appears in A.D. Crellin, "Public Cleansing", *Surveyor Local Government Technology,* 133(4010):79–80, April 19, 1969.

[s]See J. Fowler, "Truck Collection No Longer Is 'Acceptable' for City Refuse", *Compost Science,* 9(2):12–15, Summer 1968.

from the residence or building. A continuous low velocity air flow is maintained in the pipe to prevent dust or odors from escaping from the disposal chute.

Possible hazards of such a system were recognized in the study as: (1) possibility of children or pets climbing into the chute, (2) damage from fires in the chute, and (3) odors or insects coming into the home from the chute. Careful construction and placement of the chute doors would prevent personal accidents, while smoke and heat detectors would decrease fire and fume hazards by initiating water sprays and immediate collection. The scrubbing action of the in-motion waste coupled with daily collection, would clean the system and minimize insect larvae growth. The continuous low pressure suction would be combined with a sanitizing spray to further facilitate cleanliness. One hazard not mentioned in the study is the possibility that highly explosive or combustible materials could be thrown into the system. These could damage valves, pipes, and pumps and create further problems. Bulky items that would fit into the chutes could also present some problems. A degree of education of the resident could help prevent failures caused by system misuse.

Erosion of collection pipes must also be considered in the design of such collection apparatus. The ideal pipe for such a waste system would be low priced, easy to handle and install, have a smooth friction-free interior surface, would not be so porous as to cause pressure loss, would not be subject to fracture by earth movement, and would not erode from physical or chemical action. The practical pipe would obviously incorporate tradeoffs among these restraining factors.

As would be expected, the costs of collecting solid waste with an underground, completely enclosed pneumatic system are lowest in areas of very high population density. In suburban neighborhoods, where population densities are low, the cost of installing greater lengths of pipe to service fewer residents is prohibitive. In the VaCol study, the costs in dollars per ton of solid waste collected vary from a low of $6.71 per ton for a high density neighborhood to a high of $31.90 per ton for a low density neighborhood. While the economics of a pneumatic system could possibly be justified in some regions if its social value is considered to be advantageous, the costs are still high compared to truck collection. As the pipe must be buried, pneumatic systems favor areas of new construction rather than established regions where curbs, streets, sidewalks, and other services would have to be disturbed by installation.

Based upon studies under the Department of Health, Education and Welfare's Operation Breakthrough, improved solid waste systems in the collection, handling, and storage of refuse are required in new housing demonstration projects. A performance specification for a pneumatic system has been prepared and may well serve as a model for use in other urban systems.[t]

on site process

On-Site Processing of Waste

On-site processing of solid waste is another broad area of concern. Commercial, industrial, and residential wastes generated on-site are usually within the control of the producer. Public waste, however, presents a rather unique problem. This latter category includes abandoned vehicles, street refuse and litter, vegetation from parks and parkways, street sweepings, and some construction and demolition waste, as well as material from vacant lots. On-site processing of public waste is usually minimal and the material is most frequently picked up and transferred directly to the disposal site. One of the few processing functions performed in the public sector, other than collection truck compaction which is discussed in Chapter 4, involves leaf and shrub mulching. Vacuum pick-up followed by shredding, chipping, hogging, and either compacting or baling are now common municipal processing functions.[43] Several companies have suggested that collection trucks incorporating a vacuum system could be used in low population density areas where costs of connecting all homes to a pipeline are prohibitive. In this application, each home would have a disposal chute terminating at the curb. The collection truck, similar to those presently used to vacuum streets, clean sewers, and pump industrial sumps, would attach its hose to the curbside terminus of the chute. The solid waste would be pumped into the truck eliminating trash cans and crew handling, while retaining the desirable aspects of a closed system. Available equipment and information on such municipal equipment costs and performance[u] are annually updated at industrial shows and expositions.[44]

Residential processing is primarily limited today to minimal presorting or segregation of refuse constituents, and to an extent, the compaction of refuse to reduce volume in storage. Presorting may be nothing more than the emptying of

---

[t]"Solid Waste Management in Residential Complexes", Greenleaf/Telesca, Planners, Engineers and Architects, Environmental Protection Agency Report SW-35c, 1971, Appendix F, p. A-70 – A-105.

[u]A particularly complete reference regarding technology, equipments, and index to advertisers appears as *1973 Sanitation Industry Yearbook,* (10th Edition) available from Solid Waste Management Magazine, 150 East 52nd Street, New York, New York 10022. Price $20, Circulation 20,000.

separate waste cans into a large barrel or sack, or it may consist of rather extensive sorting, baling, or compaction of the waste. It is quite possible that one of the greatest advancements in collection of solid waste could be made in on-site sorting at the source. In an ecology conscious society, the presorting and packaging by the householder, commercial establishment, or manufacturing firm of various materials found in solid waste could simplify the tasks of recycling.[45] This preprocessing could also make scavenging an economically practical means of recycling much of the material and could assist in conserving many natural resources.

Presorting of waste at the source has been carried out to one degree or another by some households for years. Separation of newsprint from other trash is probably the most common practice. By stacking and tying newspapers, they can be removed for contribution to various church, service, or school groups, where the paper is sold to scrap dealers. Returnable bottles are often the subject of collection campaigns by neighborhood youth. Most recently, the separation of aluminum beverage cans has been promoted and supported by various organizations. Appliances, furniture, and clothing with reuse potential are often segregated for pick-up by—or delivery to—organizations assisting needy families.

Some commercial establishments encourage the stacking of collapsed cardboard boxes for pick-up by scavengers or return to suppliers promoting recycling programs. Manufacturing facilities are frequently in a position to separate much of their scrap for reuse in their own processing or for reselling to a raw materials supplier.

The domestic producer of solid waste could conceivably play a more important role in the collection process by placing greater emphasis on prehandling of household trash.[46] However, this would require some effort on the part of the individual citizen. The separation of glass, aluminum, steel cans, plastic, paper, and other materials would facilitate segregated pickup by municipal or private haulers, or possibly by independent scavengers. Such presorting would eliminate the formidable task of attempting to segregate mixed trash at a major collection terminal.[V] There are, of course, disadvantages to the householder in such presorting. Different containers would be required. The aesthetics of handling refuse in the home is not particularly appealing. More

[V]The general population, except in wartime national effort, does not voluntarily segregate refuse. The American public must come to realize that recycling of waste products and an improved national economy and environment begins at home. See "Reuse and Recycle of Wastes," Proceedings, 3rd Annual Northeastern Regional Antipollution Conference, University of Rhode Island, July 21-23, 1970, Technomics Publishing Co., Stamford, Conn., 243 p.

space is required for sorting the various materials awaiting collection. It is also difficult to separate products made of a combination of materials, such as bimetal cans.

Some municipal and private collectors stipulate certain presorting be performed before the trash is considered acceptable for pick-up. This is especially true if garbage is not handled by in-the-home disposals. Where garbage is collected by some service, there may be requirements that the householder drain the excess moisture and wrap the foodstuff before depositing it in a water tight container. This is advantageous to the resident, as well as the collector, because it helps eliminate odors and discourages flies, domestic pets, and rodents.

Placement of the trash on collection day also falls under the jurisdiction of the collector.[47] Some services require that refuse be placed at curbside. In some cities, the placement of trash at the curb more than 24 hours prior to pick-up is an offense punishable by fine.

There are limits on the size and weight of items that will be picked up by many services. Thus, the responsibility for removal of old large appliances, furniture, or other heavy, bulky, nonstandard trash may fall upon the individual. Arrangement must be made with the hauler to collect these items. In cases where the resident places large items at curbside, they will probably be left by the collector. A citation or fine may be the consequence of extended curbside abandonment of such trash.

Residential processing of wastes may occur through utilization of the sink garbage grinder. The home garbage grinder has effectively replaced a small part of the function of the trash can by providing a rapid and efficient means of converting foodstuff to a water-born "slurry" that can be handled by a sanitary sewer system. However, the garbage disposer is limited to just food items. While larger, more powerful trash grinders have been developed that can grind almost all nonbulky waste, their costs are high and they are not compatible with already overloaded sanitary sewage systems or home use. Multifamily dwellings and institutional establishments such as hospitals have employed these large grinders or wet pulverizers, but their reliability and performance has not yet been practically proven.[48] More research is needed in this area before any large-scale installations can be seriously considered.

Yet another mode of residential processing is demonstrated by the domestic compactor.[49] These devices are sold as kitchen appliances and can safely be operated by the housewife. The compacted trash is deposited in a plastic bag that measures 9 by 16 by 18½ inches (or about 1.5 cubic feet) when filled. Depending upon the material being compacted, the filled bag will weigh between 20 and 50 pounds. It is possible that one bag from the compactor could hold the equivalent trash of two 32 gallon garbage cans.[W] When considering that the capacity of these two cans is approximately 8.5 cubic feet, the volume reduction is impressive.

Compactors for domestic use have their advantages and disadvantages. One advantage is that trash being stored prior to collection requires less space. However, it still requires some intermediate container, such as a closed trash can or heavier bags, as the bags used with the compactors are not completely odor and drip proof. Another advantage is that the frequency of trash pickup could be decreased if all residents of an area used such compactors.

A disadvantage of a domestic compactor is that the density of trash is increased to the point where a larger container filled with bags of compressed trash could become unmanageable by collection personnel. But, the principal disadvantage that forestalls the widespread use of the domestic compactor is its high capital cost for the residence.

A novel approach to residential waste processing is suggested by a Swedish system.[50] The concept employs a refuse chute from the kitchen and toilet, being independent of the water and sewage networks, connected to a plastic decomposition chamber where wastes are biologically decomposed during long-term storage. This system does not appear practical for urban U.S. application.

Industrial operations are especially important in the preparation and on-site processing of reusable waste. The scrap waste from a manufacturing plant is of known quality and predictable volume. If the manufacturer cannot recycle the material within the plant, the waste can usually be sold to the materials source where it was purchased. Valuable scrap from a plant seldom becomes a burden on the municipal or private collector, as enlightened industry management recognizes its cash potential.

Baling of industrial waste is appropriate in plants where the scrap has some resale value. This is especially true where material is cut or trimmed to size. Quantities of waste of a known quality and content are produced and a compact bale simplifies handling while reducing volume.[51]

The primary processing mode in use on-site is the compaction of refuse as done in residential apartments, and commercial and small industrial structures. [52] These on-site refuse compactors are sometimes incorporated as an integral part of the trash container, or truck body, or they may be separate devices. Compactors have several advantages in such buildings. They can reduce the volume of the waste, with subsequent saving in valuable storage space, and compaction can assist in preventing the scattering of loose trash. If sufficient pressure is used in the compaction process, refuse will resist rodent intrusion. Also, the

[W] "Mash Trash in the Kitchen", Consumer Reports, January 1971, p. 48.

compacted refuse may be easier to handle allowing longer times between pickups.

A very large family of equipment and devices exist to accomplish the on-site processing tasks. When recalling that processing is defined as a change in physical or chemical form or removal through segregation of constituent waste categories, the number of variations in processing on-site are apparently quite extensive; the principal types and their characteristics are presented herein.[53]

The stationary packer cart is an ancillary equipment but serves an integral part of compactor processing in addition to its storage function. Although known as a cart, this device is a heavy steel box which requires the use of a large special dumping device which, in turn, is attached to a large capacity stationary packer. The cart is castered and may be drawn or pushed manually or by some type of towing unit. It can be coupled with similar carts and thus made into a train. They are made in sizes ranging from three to five cubic yard capacity, but usually the three or four cubic yard sizes are used with a five cubic yard stationary packer. The three or four cubic yard carts cost approximately $450. In use, these containers require a special type of dumping device after staging them on a platform either at ground level or at truck height. The dumper lifts and tilts the container, which is securely latched onto the dumper platform, and empties it into the charging box of a stationary packer. The ground level model of this dumper costs approximately $2,500. The dock type dumper requires a special ramp and costs approximately $2,100.

The open topped roll-off container is also in versatile use associated with compactors that precompact wastes for storage in these containers. Built of heavy steel, these containers are designed for the mechanical dumping of loose wastes and very large noncompactable objects. They are frequently used in connection with self-dumping hoppers handled by fork lifts.

Capacities of these containers range from six to forty or more cubic yards. Over-all widths are about eight feet. Heights vary from three to eight feet. Lengths run between seventeen to twenty-one feet. These big boxes are handled onto and off truck chassis by tiltframe hoists of various designs. Dumping the contents of the container is accomplished by raising the box and opening its rear doors. Prices range from $400 for ten cubic yard to $800 for twenty cubic yard units.

The most common type of container in use in large buildings is an integral part of the stationary compactor. These containers, generally box-like in appearance, usually have an opening in the lower half of one end of the box to allow the compactor ram to operate inside the container during the compaction cycle. This loading end is also hinged as a tailgate to permit it to swing fully open for the dumping of refuse from the container to the collection vehicle. The container is equipped with a ratchet locking device used to secure the container to the packer during the filling and compacting operations. When the container is full, the ram has been entirely withdrawn and the retaining cables are in place, the full container is winched onto a tiltframe hoist and can be carted away or dumped into a collection vehicle—depending upon the container's size. The full container is replaced by an empty one, which is strapped or locked to the packer and the entire unit is again ready for operation. The sides and top of the container are tapered slightly to facilitate refuse slide-out when dumping. The following data indicates the price range of the various sizes of large tiltframe containers available:[x]

Capacity (cubic yards)	*Length (feet)*	*Approximate Price*
27	16	$3,000
30	18	3,100
37	20	3,400
42	22	3,500

The stationary compactors in use cover a broad range of size and capability. [54] Variations among many types include those with horizontal and vertical rams, hydraulic or pneumatic actuation, manual or automatic or semiautomatic control, single bag and continuous or multibag units, and accessories such as conveyors, carousels, and billeting equipments for compacted refuse removal to storage.

Bag-type compactors can be chute-fed, and manufacturers claim productive capacities of equipment ranging from 7 to 44 cubic yards per hour and compaction ratios ranging from 3:1 to 12:1. Single bags must be removed when full and replaced by empty ones; continuous multibags must be tied off, removed, and replaced; filled containers must be replaced, all of which emphasizes the necessity for matching equipment to anticipated daily volumes and the avilability of trained maintenance and operations personnel.

Paper, polyethylene plastic, and butyl bags of the "sausage link" type are available. These are made with approximately 10 "links" and are preformed in such a manner as to permit longitudinal expansion as the links are filled. The machine extrudes the compressed wastes in slugs into the attached string of bags. The bags can be conveniently tied off, cut apart, and thus become individual units which can be conveniently handled.

Single bag stationary compactors range from $3500 to $5500 while the

[x] The source for this data was "Solid Waste Management in Residential Complexes", Greenleaf/Telesca, Planners, Engineers and Architects, Environmental Protection Agency Report SW-35c, 1971, Appendix F, p. A-70 – A-105.

multibag units cost from $4700 to 10,000. Compaction ratios from 4:1 to 8:1 and package densities from 18 to 60 pounds per cubic foot are commonly claimed by manufacturers, depending upon the composition and mix of solid wastes which will vary over a wide range. Based upon a density of 6 pounds per cubic foot for residential wastes and a realistic compaction ratio of 3 to 4 to 1, a density range of 18 to 24 pounds per cubic foot may be expected. Containerized packages weighing as much as 200 pounds, as claimed for one model, present handling problems which may require the use of more than one man for their removal and transport.

A variation of conventionally designed bag packers has a duplex ram. A smaller diameter, independently operated ram is built into the center of the main ram. Three hydraulic pistons actuate the rams—two for the main ram and one for the secondary ram. The device has a cone on its output snout. The primary ram compresses the waste materials into the larger portion of the compression cone, after which operation the secondary ram is actuated, thus further compacting the refuse and forcing the compacted materials into a receiving bag or can.

The console compactor class of equipment employs a vertical compacting ram, which may be either mechanically, hydraulically, or air operated and is usually hand-fed. Chute-fed models are in the development and testing stage. These units compress waste into a corrugated box container or a plastic or paper bag.

Models are available to process one container or two containers side-by-side, but housed within the same cabinet. In-line type compactors are also available, some of which will accommodate up to eight containers, but these are not within an enclosed cabinet and are not intended for operation by building tenants as are the two previously mentioned. The containerized packages are about 3½ cubic feet in volume but some models produce packages of from 5 to 6 cubic feet. The densities of containerized packages range between 12 and 30 pounds per cubic foot.

Claimed compaction ratios run as high as ten to one. One type provides for the suspension of the receiving bags from special holders mounted on a castered cart. This permits the filled bags to be transported within a building without need for a separate cart.

The claimed weight of containerized packages ranges from 40 to 120 pounds. It is estimated that a single unit would handle the daily wastes of up to 40 persons (based on a per capita production rate of 3 pounds per day) before a full bag or container would have to be removed from the machine and be replaced with an empty one. The capacity of the unit is limited by the frequency of service furnished by maintenance personnel. This type of compactor can be chute-fed, but such adaptation generally is considered practical for only low density housing because of the small capacity of this type of machine.

Costs for console compactors range from $1000 for single bag units to $4000 for two bag models.

The rotary compactor, usually called a carousel type, consists of a ram mechanism which packs loose wastes into paper or plastic bags held in open positions on a rotating platform. When the bag directly under the packing ram is filled to a predetermined depth, the platform indexes one position, thus moving the full bag from under the ram and replacing it with an empty one. The bags are held in place within a compartment which confines the bag and prevents it from rupture during the packaging cycles. These compactors are made in standard models of 8 or 10 bag compartments but are available to accommodate 20 or 30 bags. Originally of Swedish design and manufacture, they are now being made and marketed in the United States.

The advantages of multibag machines are obvious. Each bag will hold about 3½ cubic feet of compacted wastes, estimated by the manufacturer to have a density of up to 20 pounds per cubic foot. It is estimated that a ten bag compactor can handle the refuse generated by 200 to 250 persons for each servicing. Removal and replacement of bags can be accomplished in less than an hour by one man.

Carousel compactors having at least eight-bag capacities can be used in multistory low density residential structures. They can be chute-fed and can readily be equipped with optical or sonic controls. Costs for carousel units range from $4700 to $9300 and to $15,900 for dual units supplying a large carousel.

The stationary compactor is quite frequently termed a stationary packer. Both manufacturers and users employ the terms interchangeably. In its fairly standardized form it is a compaction unit having a hydraulically operated ram which moves in a horizontal direction. Wastes are fed into a receiving hopper and the ram, when actuated either manually or by optical or sonic devices, compresses the wastes into a steel container which, although an important part of the equipment, is a separate component that can be easily attached to or detached from the packer mechanism. The filled container, if small, can be moved by hand on its casters. The larger capacity containers must be handled mechanically by special types of equipment as discussed previously.

The stationary packer is a proven type of solid waste processing equipment.[55] An example is shown in Figure 3-1. It is capable of reducing wastes to a level of 20–25 percent, or less, of their loose volume. Charging box capacities of these packers range from about one third of a cubic yard to several yards. Packer capacity is rated in cubic yards per hour and is dependent upon charging box capacity and cycling time. Chute feeding is a common practice when these packers are used. Optical and sonic controls are available to provide automatic operation, which requires only periodic attention of maintenance personnel.

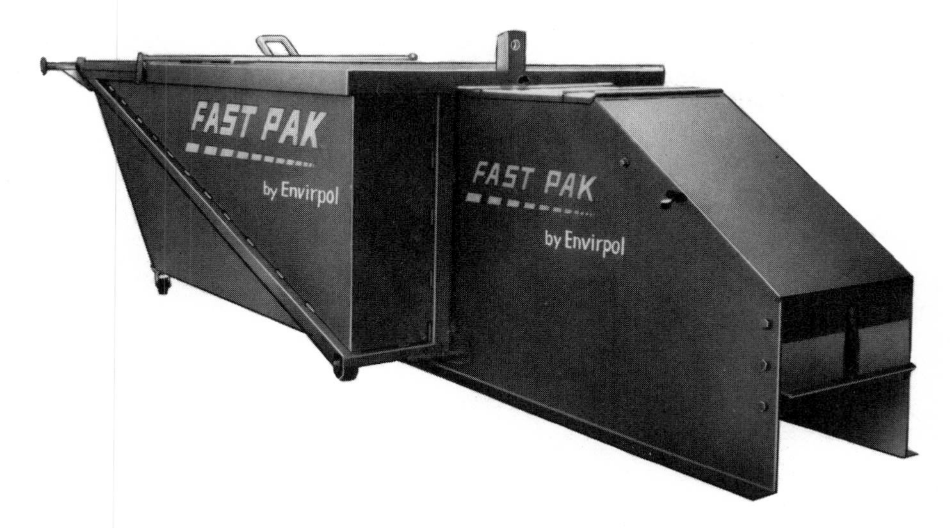

Figure 3-1. The Stationary Packer. Source: Courtesy of Environmental Pollution Research Corporation, New Hyde Park, New York.

This type of compactor, because of its large container capacity, requires less frequent attention by building maintenance personnel. The smaller sizes can be fitted with containers varying from 2 to 3 cubic yards capacity upward. This improves maximum capacities from 50 percent to 100 percent over bag type compactors. Attendants' time is correspondingly reduced.

Prices of these compactors range between $2800 and $6000 depending upon size. Containers cost from about $300 for the two-yard size to over $1000 for the ten-yard size.[y]

Bottle and can crushers and other dry grinder equipments are readily available but are quite specialized in their applications. Crushers reduce volume by as much as a ten to one ratio and cost from $900 to $3500 for the large industrial units with conveyor attachments.

Dry grinders of various styles also find greatest application in industrial waste processing. This category of equipment includes hoggers, pulverizers, hammermills, and shredders and can achieve as much as a 15:1 volume reduction ratio depending upon waste input. Costs for these types of equipments usually

[y] As displayed in Reference 56, a model contract for fixed packers may be quite useful in selecting and acquiring a compactor for building installation. A model performance specification is shown as Appendix G, page A-106 – A-121 of the Environmental Protection Agency Report SW-35c entitled "Solid Waste Management in Residental Complexes", 1971.

range from $7000 to $20,000; but these equipments can be made quite large, with a cost in excess of $100,000.

There are several solid waste processing systems on the market and in operation which utilize wet pulpers as the principal means of volume reduction. Generally, pulpers consist of a pulping bowl with a pulping impeller and a waste sizing ring in the bottom. Accessory equipment includes a junk ejector and a dewatering press. The pulper and junk ejector are mounted directly adjacent to each other but the dewatering press may be located at some distance from the pulper and connected to it by piping. It is possible to utilize multiple pulping stations and one dewatering press, and, in general, units can be located in the most convenient places, since the slurry goes to the press and water is returned to the pulper by pipelines.[57] Wastes can be introduced into the pulper by chute in-floor models, manually fed into pit models, or carried by belt conveyor into these or other models. Capacities of pulpers vary from about ½ to 2 tons of waste per hour for large installations.

In general, only food wastes, paper, and light residential wastes are being processed. Considerable difficulty has been encountered in the satisfactory handling of plastics, especially polyvinyl-chloride containers, plastic tubing, textile products, such as mops, and some of the occasional heavier materials which are found in unselected wastes. Water requirements for pulpers are based generally on 20 to 25 gallons per 100 pounds of dry solid wastes processed. Through the use of dewatering devices, the majority of water used is reclaimed and recycled in the pulping operation. It is not possible to obtain cost figures on this equipment since an appreciable portion of such costs is for installation. However, including the varying costs of installation, the total costs are estimated to range from as low as $25,000 to as much as $90,000 for a 1000 to 4000 pound per hour installation.

4 Transport Concerns

To most participants, bystanders, or students of solid waste affairs—be they government officials, laypeople, or scientific in their orientation—the transport function is the heart of the collection system. Indeed, although the definitions for "collection" are skewed in this book to include all of the concerns from generation through predisposal, the collecting of solid waste and the transporting of that refuse from the site of its conception to the site of its demise is the central aspect of waste management.

This is the place where private and personal affairs interface with government and business and in not too exposed a fashion. How can a householder feel too embarrassed over a can or sack of garbage when everyone else's mess is on display at curbside or in view when being dumped into the collection truck? In the following few pages, the presentation is offered that directs attention to this transport function—the tasks that are usually associated with collection of refuse. But the examination is to be detailed to the extent that the state-of-the art is covered regarding personnel safety and salaries, equipments utilized, management techniques employed, public and private responsibilities for conducting the tasks, costs involved with this phase of waste management, and innovations that demonstrate successes now developed by solid waste management people and industry.

Personnel Matters

Refuse collection is a labor intensive funtion and wages account for the largest expense item in the budget of both public and private collection.[1] With steadily rising labor rates, the cost squeeze is more severe than ever. The American Public Works Association sees these wages as 60 to 70 percent of the cost of refuse collection. In a breakdown of collection and disposal costs shown in Figure 4-1 and published by the Battelle Memorial Institute,[a] collection manpower is listed as consuming 71.2 percent of expenditures. To this portion is

[a]C.J. Lyons and D.L. Morrison, "Solid Waste 1: Where Does It All Come From?", *Battelle Research Outlook,* 3(3), 1971.

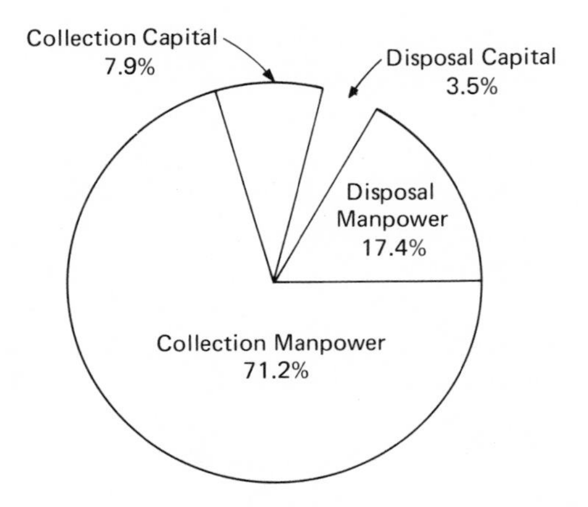

Figure 4-1. A Breakdown of Costs to Communities for Refuse Handling. Source: C.J. Lyons and D.L. Morrison, "Solid Waste 1: Where Does It All Come From?" *Battelle Research Outlook*, 3 (3), 1971, p. 6.

added 17.4 percent of the total spent for disposal manpower, 3.5 percent spent for disposal land and equipment capital costs,[b] and 7.9 percent spent for collection equipment capital investment. It is estimated that from $1.5 to $4.5 billion a year is spent collecting municipal refuse in the United States. Of the total waste collected, 90 percent is picked up and moved by hand at least once after its generation and before its ultimate disposal.

Refuse handling during collection is taken for granted by the general public. Garbage collectors, trash men, or junk collectors are low on the social order of most of our society. The working conditions of solid waste collectors are far from ideal; health hazards are high due to exposure to refuse; exposure to all types of climatic conditions during work performance is necessary; great physical strength is required of employees; there is little mental stimulation, accident hazards are numerous; and pay scales and career advancement opportunities are minimal.[2] The sanitation worker's viewpoint [3] is only experienced publically during the time of a strike and then has reduced impact due to the system's users practical plight of refuse removal. At such times of labor problems and strikes, disruptions of sanitation service are politically significant in that the

[b]Every city or private collection agency experiences a different cost meld and reports costs differently. The city of Baltimore, for example, expends $11,612,750 or 81.9 percent of its 71-72 fiscal year budget on collection labor, street cleaning, equipment maintenance, and repair and garage operations. Baltimore operated 300 pieces of equipment on 92 separate routes making 100,000 stops daily.

populace sees such problems as a breakdown of city management or another labor/management squabble to which the public is an unwilling and offended party. [4] Indeed, the status of solid waste collection personnel working in a municipally operated system differs little from that of personnel working for a private hauler, or those working for a hauler on contract to the city. While the collection personnel in a public system may benefit from civil service status, those working for private or contract haulers do not enjoy this advantage.[c] Many sanitation workers have seen advantages in unionization or collective bargaining and extensive, crippling strikes by workers have been carried out in various cities in the last five years.

A 1970 study by the U.S. Department of Labor indicated that wage increases for union drivers and collectors in solid waste operations averaged 5.9 and 5.8 percent respectively between July 1968 and July of 1969. As of July 1969, drivers were being paid $4.01 per hour on the average, while collectors were paid $3.65 per hour. Increases over the 1968–1969 period were the highest since the study began in 1936. [6] As would be expected, the range of wages varied from city to city. Some major cities paying the highest wages were Spokane, Seattle, San Francisco, San Diego, Portland, and Los Angeles, where drivers make $5 or more per hour. [7] Drivers in many cities earned less than the $4.01 per hour average; some in Knoxville, Little Rock, Memphis, New Orleans, Phoenix, Toledo, and Shreveport were paid less than $3.50 per hour.[d]

Hiring and Training

The hiring of solid waste collectors and drivers presents one of the more important problems a municipal or private service will encounter. The "garbage collector" is considered by most people to be at the very bottom of the socio-economic labor ladder and the available positions hold little appeal to those seeking work, even in times of high general unemployment. A better image of such work has become part of the sanitation program public relations effort in

[c]The San Francisco sanitation system is a private entity that overcomes many potential labor management problems because it is employee owned. As described in Reference 5, collection cost to subscribers is $2.50 per family per month from Sunset Scavanger Company. Approximately 250 of Sunset's 500 employees own 32 shares each of the company which began as a cooperative between the numerous so-called "Ma & Pa Operations".

[d]The range of wages as documented by the Department of Labor indicate for drivers of collection trucks a high of $6.00 (Seattle) and a low of $2.50 (Toledo, Ohio); for collection men or truck helpers a high of $3.95 (Des Moines, Iowa) and a low of $2.85 (Boston). Many jurisdictions and firms make no distinction in these job assignments other than that based on seniority and promotion.

many cities. Classified help-wanted advertisements generally call for "sanitation service men," or some other title that is not distasteful. Some municipalities supply their collectors and drivers with uniforms that resemble those of law enforcement officers. Health insurance, retirement, and quota based incentive programs are also used to encourage applicants.

Several considerations may be made to hire and retain a capable staff of waste collection personnel, whether for private or municipal employment. These matters include:

1. Improved wage structure–Wage structuring is not wage escalation. Employees within an organization and a geographic region need to know that labor is uniformly expected and rewarded.[8] A rigorous organization structure informs the employee of the supervisor's expectations, as well as the relationship of other supervisor personnel who may be making requests of the employee. Job descriptions give the supervisor a bench mark for rating the employee and allow the employee to occasionally rate himself. Organizational charts also help to show the dedicated employee paths for possible advancement within the organization. The salary ranges and compensations for job advancement must be clearly delineated to employees.[e]

2. Selection of applicants–Screening available applicants for collection labor is not a simple process of selecting someone smart enough to carry a trash can. Slow reaction, inability to follow instructions, limited physical capability, poor eyesight–all could endanger the worker. Physical fitness certainly is a major consideration when hiring, as those with a poor medical history or with physical problems uncovered in a medical examination have poor probability of retention by the employer. Preemployment and annual physical examinations are highly recommended as criteria for employee selection. A past history of work habits, undependability, frequent absences, and poor moral fiber should be considered as cause to disqualify the applicant. While few overqualified individuals will apply for such positions, the personnel manager should guard against hiring underqualified persons just because they are available and will accept the work.

Drivers must fulfill more requirements than assistants or laborers and should be screened even more critically. A chauffeur or truck driver license is required for such jobs, and holding of such a permit is one indication of capability. The driver must also be insurable. Drivers are usually the crew chiefs or supervisors of the groups.

3. Training–Training of new employees is necessary to familiarize them

---

[e]The City of Cincinnati has a model organization layout to serve as a guide and reference. See "Refuse Collection Practice", Public Administration Service, 1966, by American Public Works Association.

with procedures in the system as well as routes.[9] Safety instruction is mandatory. A portion of the training should be supervised work in practice and it is during this period that individual problems or incapabilities can be detected and corrected. The training process should also be used to sell the employee on the importance of his job to the proper functioning of the city in which he is working. Training should be formalized so as to ensure it is a part of the new employee matriculation. As new equipment is purchased, route changes are made, or other innovations occur, supplementary training may be required.[10]

In order to acquire and retain efficient and effective collection crews, the health and safety of the personnel should be carefully monitored and stressed in job training activities.[f] Some waste collection routes will be subject to handling of volatile or dangerous wastes and instruction in the proper transfer of such materials is an absolute necessity. Chemically oriented industries, metal fabricating plants, biological laboratories, hospitals, and similar sources of wastes will presumably notify the waste collectors of hazards in their refuse. If they do not, it is the responsibility of the collection system management to acquire this information and inform the crews with appropriate training assistance. Equipment safety [13] and operating procedures [14] to reduce accident rates and lost time can result in dramatic profitability improvement in the case of private haulers or substantially reduced costs and tax burden in the public operating sector.[g]

Health and Safety

Health of sanitation workers is also among the most jeopardized of any worker category. Incidence of disease is correlated to the work environment [16] where infectious viral and bacterial presence causes lost time medical disability at a level four times the national industrial average.[17]

Personnel health and safety hazards encountered by refuse crews have been listed in summary by the National Safety Council as occupational hazards to be curtailed. These include:

1. Excessive haste—which may be a result of the crews attempting to qualify

[f] In California the estimated injury rate for privately operated refuse collection vehicle employees was 180.9 disabling injuries per 1000 employees, nearly six times the average for all California industries (see Reference 11). Reference 12 indicates that solid waste employees throughout the nation experience nearly 900 percent greater incidence rate of injury than all U.S. industries.

[g] As discussed in Reference 15, reduction of accident rates could double the profitability of numerous private haulers.

for some incentive program. In systems where the crew is allowed to go home after completing their route, there will be an incentive to hurry. A similar incentive may exist where there is a premium for volumes of refuse exceeding a preset amount.

2. Stepping on or off trucks—especially when they are in motion is a dangerous practice commonly encountered in waste collection.

3. Lifting—which can include lifting of containers which may have been overloaded by the resident or improper procedure in lifting. Containers which are improperly shaped for easy lifting contribute to such accidents. Containers may lack handles, or have broken handles, or they may have bottoms weakened by rust, or in the case of paper containers, they may have been exposed to water which has weakened their structure.

4. Yard hazards—which include garden hose, lawn tools, toys, construction material, fences, and other obstacles to access.

5. Handling hazardous objects—such as broken glass, sharp metal objects, splintered wood, fluorescent lights, or chemicals such as weed killers or insecticides.

6. Poor access to trash—through narrow alleys, between buildings, up or down stairs, or from treacherous storage areas. Access walks may be covered with ice, oil, or other materials.

7. Pedestrian accidents—especially when collecting on heavily traveled thoroughfares, during high traffic periods, or stepping from behind a vehicle to cross the street. Solid waste collection crew members spend more time on the street than any other worker.

8. Animal and insect attacks—dogs, rats, wasps, spiders, and other creatures threaten the collector.

9. Exposure to the elements, dust, fumes, infected material, disease—may not result in an immediate reaction, but results may be cumulative or sustained. Conjunctivitis (pink eye) and dermatitis are hazards of collectors.

10. Mechanically caused injuries—such as fractures, cuts, bruises, and sprains from defective equipment, or improperly operated equipment. Tailgates, doors, compactors, hydraulically actuated devices, and other mechanical systems are hazards. Sweeping blades on compactors are extremely dangerous and the National Safety Council makes note of injuries occurring while collectors are attempting to retrieve salvageable objects from such mechanisms.

11. Fatigue, personal problems, daydreaming—coupled with any one of the previous hazards can noticeably increase the possibility of injury. Most accidents are still failures of the human system.

Federal law requires job connected accidents to be reported and the individual solid waste collection system will undoubtedly maintain complete records of mishaps for insurance purposes. Periodic review of accident reports should be used to detect problem areas.

Efficiency of Personnel

Various aspects of personnel quotas and efficiency have received a great deal of attention lately due to the recognition that the collection of refuse is labor cost oriented and tools are now available to alleviate problems associated with those high costs. The tools used are management decision making abilities aided by use of computers. [18, 19] These personnel/equipment interfaces may be reviewed from the viewpoints of efficiency in route planning, collection scheduling, collection crew size, and efficiencies promoted through competitively applied quotas for operating personnel.

Work quotas or work loads in solid waste collection are measured in man minutes per ton or man minutes per yard of refuse. There are no standards for such service as the variables affecting the efficiency of crews cover such a wide range. Some of these variables are:

1. *Density of population along route*—Apartment complexes with centralized trash pickup areas provide solid waste in a form requiring little more than loading into the collection vehicle. On-site compactors are occasionally found in such developments and these offer high density solid waste. In residential areas where homes are built on large lots, the collection crews must spend more time in transit.[20]

2. *Placement of refuse and crew size*—If the crews have to go through yards, into garages or basements, or spend time walking between trucks and trash storage areas, the efficiency will be lower.[21] Curbside pickup decreases wasted time and effort.[22]

3. *Type of containers used*—Bagged trash eliminates the procedure of returning cans.[23] The collector has only to throw bags into the pickup vehicle. Trash that has been compacted in a home unit increases the efficiency by raising the load capacity of the vehicle. If the waste must be shovelled from some enclosed vault into the truck or auxiliary transfer container, the time will be excessively long.

4. *Type of vehicle used for collection*—Open vehicles, with high sides, and no integral compactor are more difficult to load and must make more runs with less dense loads to the landfills or transfer station. A large compactor truck eliminates a portion of these problems.

5. *Vehicle transit time*—If long trips to the disposal are required, and the crews ride with the trucks, the efficiency drops.[24] Ideally, another collection

truck should take over the route and the pickup crews as soon as one vehicle is loaded.

6. *Weather conditions and seasonal phenomena*—Snow, rain, cold, and extreme heat affect crew efficiency. Quantities of vegetation refuse during the summer increase the work load and route time. Holidays, shopping days, and other variables cause fluctuations in the amount of waste to be collected.

7. *Route planning*—The development of planning to minimize the length of a solid waste route for a specified collection area will increase overall efficiency. [25, 26] Collection from both sides of the street during a single pass, minimizing the number of left turns onto busy thoroughfares, and maximizing the loading per stop are some important aspects of route planning.[h]

The Public Health Service Bulletin No. 1892, previously cited as Reference 21, contains significant conclusions regarding the personnel efficiencies and equipments used in the waste collecting function. These conclusions are itemized below.[i] Items 1 through 11 are based upon field studies and MTM analyses; items 12 through 14 are based upon national survey data,[29] and items 15 through 23 are conclusions drawn independently by a contractor.

1. For curbside collection of refuse, one-man crews were more efficient than multiman crews; the productivity of the one-man crew was greater than that of the multiman crew when measured in terms of route man-hours per ton.[30]

2. The one-man crew was similarly more efficient than the multiman crew for alley collection of refuse.

3. Multiman crews were more efficient for backyard carryout collection of refuse.

4. Under specified assumptions for important route factors and costs of equipment and labor, the unit cost of refuse collection by the one-man crew was 25 to 45 percent less than the two-man crew and 35 to 50 percent less than the three-man crew.[31]

5. Although multiman crews required less equipment of equal size than the one-man crews, this had a negligible effect on unit collection costs when the combined equipment operating, amortization, and labor costs were compared for one-man and multiman collection.[32]

6. In residential or light commercial curb or alley collection, the workload was not excessive for one-man operation.

---

[h]The use of method-time measurement (MTM) is a significant aid in the study of efficiency and route planning to maximize crew usage. See References 27 and 28 for examples of the application of this technique.

[i]References have been added to direct the reader to amplifying information on discrete points.

7. With existing collection equipment designs, side-loading compactor vehicles were the most suitable type of one-man operated equipment for curbside and alley refuse collection operations.

8. Significant savings in curbside collection time were achieved by the use of disposable containers such as paper or plastic bags.[33]

9. Industrial time standards developed for production control and design were found applicable to evaluating the task of refuse collection.

10. Based on preliminary human factors studies, the weight of the refuse container and contents was more important in the rate of collection personnel performance degradation than vehicular loading height.

11. Various complex refuse collection system interrelations affect optimum crew size, equipment, and cost benefits.[34]

12. The predominate (1968) practice of refuse collection used by a sample of 234 cities involved the use of rear-loading packer vehicles with three-man crews for curb and/or alley collection.

13. Only a limited number of small cities used one-man crews for refuse collection.

14. In a sample of 234 cities, unit operating costs for collection of refuse generally increased with the size of the city.

15. Public refuse collection systems in general have been slower than private collection systems to adopt new refuse collection technology such as smaller crew sizes, certain low-cost or high-efficiency equipment types, and related system modifications.[35]

16. If labor costs and the incidence and severity of collection labor strikes continues to increase, the one-man collection system may become more common, particularly in private collection firms and in smaller cities.

17. Current municipal collection systems are frequently characterized by: personnel with limited skills and work experience; high absenteeism; absence of promotion opportunity; and lack of public recognition of the collection worker's contribution.[36]

18. As the cost-benefits associated with the one-man crew are sensitive to excessive absenteeism and poor work habits, the one-man collection system generally requires a higher level of responsibility, performance, and loyalty on the part of both collection and supervisory personnel.

19. Successful implementation of a one-man collection system will probably require: higher personnel standards; higher salary rates; potential upward mobility in the job structure; employees with a sense of personal pride and responsibility; and engineering evaluation of route structure and equipment requirements.

20. There is an immediate need for improvement in the design and applica-

tion of specific equipment for refuse collection tasks. The combination of packer body and conventional truck chassis does not provide for an optimum man-machine relationship.

21. Many existing collection systems can be significantly improved by engineering design of collection methodology, including crew and truck sizes. [37]

22. Increased awareness by collection system administrators concerning potential cost savings and improved human factors can lead to the demand for and use of better equipment designs. [38]

23. Careful planning and engineering of the collection system can realize maximum public health protection cost savings, improved service, and reduction in the frequency of labor strikes and other personnel difficulties.

Conversion to one-man pickup crews has been carried out in Ingelwood, California, a community formerly utilizing two-man crews with rear load packer trucks. The city now uses larger, 35 cubic yard trucks with side loading for one-man operation.[j] In comparing statistics from 1960 to 1971, the period of transition, the changes shown in Table 4-1 were noted. Of particular interest is the decrease in man-hours per ton of collected refuse. Even assuming a 5 percent per year inflation rate over the 11-year period cited, the net saving in dollars per ton of collected refuse is still 32 percent less in 1971 over 1960 for the high labor costs of collection.

Responsibility for Collection

The final economic responsibility for collection of solid waste will always lie with the generator of the refuse. Eventually, the costs of collection are paid for by the generator through taxes to the municipality or fees paid to the collector.

In some communities, the local government manages the collection system in toto. The government may legislate requirements for solid waste pick-up, own the trucks, hire the crews, and manage the disposal operation. In other communities, the actual collection and transportation of the waste may be contracted to one or more firms meeting the approval of the city. In other locations, private haulers may remove refuse on a basis that may be competitive with other haulers. Even in the latter case, where the city has little or no control over the actual collection process, local ordinances may specify the frequency of collec-

---

[j]W.F. Farnum and H.M. Frisby, "Humanizing Refuse Pick-up," *Waste Age*, September/October 1971, p. 22.

Table 4-1
Compacting Statistics for 1960-1971

	1/1/60	*1/1/71*	*Percent Increase or Decrease*
Population	66,598	95,000	+42.6
Dwelling Units	25,330	38,031	+50.1
Annual Tons of Refuse	24,265	37,531	+54.7
Total Truck Loads (or trips to landfill)	5,855	4,335	-26.0
Annual Man-Hours	52,167	36,534	-30.0
Man-Hours Per Ton	2.19	0.97	-55.7

Source: W. F. Farnum and H. M. Frisby, "Humanizing Refuse Pick-up," *Waste Age*, September/October 1971, p. 22.

tion or limit the amount of on-site trash. This type of ordinance has significant enforcement problems, though.

In general, it has been shown that municipal solid waste systems cost more to operate than private or contract systems.[39] In New York City, the private haulers cost for collection is $17.50 per ton while the city tax supported system cost is $49.00 per ton (see Reference 40). The reasons for the higher per ton costs of a municipal system may be attributed to the fact that the municipal crews are paid on a straight hourly basis and there is seldom an incentive to collect more than some predetermined quota from the route.[41] If minimums are set, they usually allow considerable freedom in comparison to those set by a private contractor operating on a competitive basis.

On the other hand, the private collection service is usually a smaller operation with an incentive to make a profit, an influencing factor that does not affect the municipal system.[42] As the private service will undoubtedly increase its profit with increased volume, the incentive to the owner is apparent; part of the profit may be passed on to the crews to enhance business stability and further growth.

The contract collector is in a position similar to that of the private operator. While usually receiving a fixed fee from the city or regional solid waste management authority, efficient use of equipment and personnel allows increase in profit. Inefficient uses of either of these resources require the addition of costly people and equipment to meet contractual commitments.

Considerable debate is waged pro and con for municipal or private responsibility. Some parties favor municipal assumption of responsibility [43] and others look to the free enterprise system.[44] Yet other commentators feel there is benefit in cooperation by the dual approach [45] and are criticised sharply by the private sector for suggesting that government compete with business.[46]

The public/private duality exists abroad,[47] exists in small geographic locales domestically,[48] and is accommodated in planning by those who favor regionalization of refuse collection responsibilities.[49] Those parties who cite success to demonstrate a particular viewpoint [50] must be heard as proponents with biased authorship. The facts are that the American Public Works Association has noticed a definite trend toward contract and private systems assuming a greater share of the total load of waste collection. Whereas total municipal systems were utilized by some 55 percent of the cities in 1955, only 45 percent of the cities in the 1964 survey had completely captive systems. Various combinations of contract (private collector under municipal contract) private, and municipal systems are being utilized (see Table 4-2).

In a 1969 publication of the Office of Science and Technology,[k] the results of a survey of 6,300 communities indicated 53 percent of the total exercised no control over on-site storage of household garbage. However, the larger communities exercised more control than the smaller ones and on a population basis, 56 percent of household inhabitants are served by public collection agencies, 32 percent by private collectors, and 12 percent handle their own waste problems.[l] Commercial institutions depend heavily on the private collector, as do industrial concerns. Figure 4-2 shows a breakdown of collection methods on a national basis. The large segment of industrial waste that is self-collected is due, in part, to sale of industrially generated scrap for reprocessing. As for manpower utilization, 192,000 people were employed in public collection systems while 145,000 worked for private collectors.

Regardless of the public or private nature of the collection agency the tasks and obligations remain the same.[51] The agency must see that collections are made on a regular basis, with minimal inconvenience to the citizen, with enforced conditions of sanitation and safety, with the smallest amount of traffic disruption, and at the lowest cost. The collector must also be assured of cooperation on the part of the citizen. This is usually achieved through ordinances enforced by the city, although reports and rectification of violations may fall under the responsibility of a contractor.

Contract specifications issued by a municipality or regional waste authority

---

[k] Solid Waste Management, prepared for Office of Science and Technology, Executive Office of the President, May 1969.

[l] No explanation is offered for this apparent reversal in the trend over the short five year period. For a pointed discussion of governmental collection inefficiencies and competition with the private sector, see "Private Sector's Ability to Pick up for 65 Percent Less Arouses New York Officials", *Solid Wastes Management/Refuse Removal Journal*, June 1971, p. 16.

Table 4–2
Type of Collection Agency Used by Cities in 1939, 1955, and 1964. Prepared by American Public Works Association

Collection Agency	*1939 Survey Number of Cities*	*%*	*1955 Survey Number of Cities*	*%*	*1964 Survey Number of Cities*	*%*
Municipal	105	55	494	55	446	45
Contract	34	18	134	15	175	18
Private	20	11	95	11	130	13
Municipal and Private	20	10	51	6	151	15
Municipal and Contract	4	2	72	8	33	3
Municipal, Contract, Private	–	–	26	3	16	2
Contract and Private	7	4	21	2	49	5
Total	190		893		995	

to contract for refuse collection services are quite inclusive.[m] When licensing or franchising of a contractor by a governmental unit for collection service is contemplated, these same contract specifications may be employed but the contract document is different.[n] In cases where several contractors elect to form a cooperative corporation to participate in a license or franchised operation, or otherwise select this joint venturing for profitable growth, another form of contract is necessitated.[o]

Waste Collection Financing

The problem of suitable financing of solid waste collection is quite basically one of acquiring sufficient funds to achieve the desired level of service, with an equitable distribution of those costs among the taxpayers.[53, 54] Fortunately, most communities have some historical base that has, over the years, been adjusted to meet past and present demands, and can be projected (with alterations) to meet future demands. Individual household services are commonly

[m]The American Public Works Association lists sixty one major considerations for such specifications in *Refuse Collection Practices,* 1966, and discusses contract methods and procedures.

[n]A sample franchise agreement is displayed in *1972 Sanitation Industry Yearbook*; see Reference 52.

[o]*Ibid.*, p. 8, "Agreement to Form Cooperative Contractor Corporation".

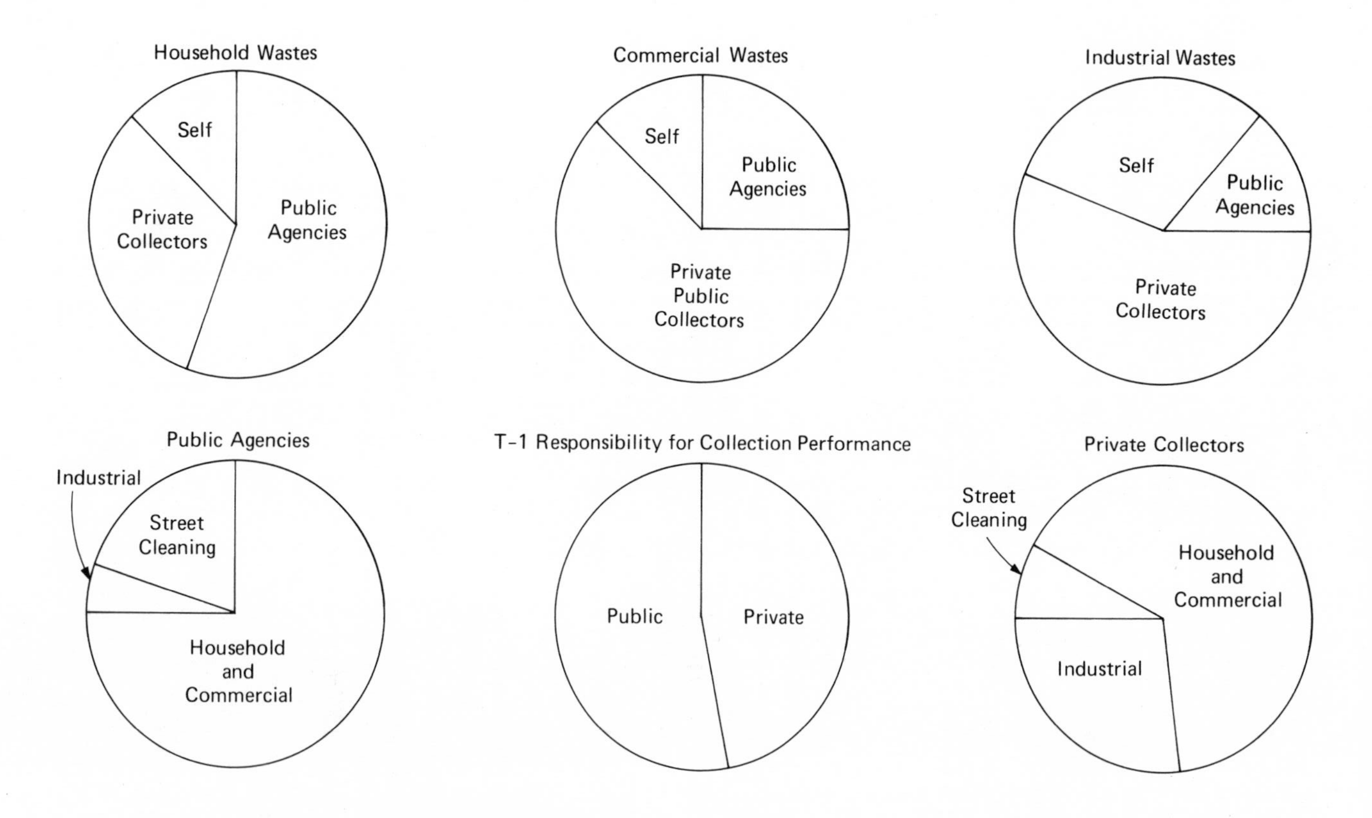

Figure 4–2. Manpower Utilization in Collection Activities.

handled by general taxation. Services that are conducted for the benefit of special interests are commonly charged to those interests on a fee basis.

In recent years, problems of financing have arisen where there have been high step increases in labor rates, where communities have annexed areas without proper services, and where improvement or modernization programs have been initiated. Often drastic increases in the cost of a solid waste management system result from a need for immediate and large-scale improvements in methods of disposing of waste—a new incinerator is required, an old landfill becomes saturated, or a new pollution standard must be met.[p] Increases in primary collection service costs are usually more gradual and easier to cope with unless some community is updating a particularly antiquated system. However, in the public sector, the costs for waste management and its capital financing needs come from a government assessed charge such as the following:

1. *General Property Taxes*—Regular appropriations are made from general revenues obtained from annual property taxation, state-collected, locally-shared sales taxes, or similar sources.

2. *Separate Property Taxes*—This involves levying separate ad valorem taxes, usually on the same basis as general property taxes, for a specific purpose such as refuse collection.

3. *Service Charges or Fees*—These are established charges made to householders and other producers of refuse, on the basis of the measured, estimated, or presumed amount of waste removed.

4. *Can and Container Rental Charges*—These are established rates or charges made to provide householder and other refuse producers with municipally owned, standard refuse cans, bulk containers or other receptacles, including paper or plastic bags, and to cover the cost of emptying, servicing, maintaining, and replacing such containers.

5. *Special Assessments*—These are similar to service charges except that assessments are made against the properties benefited, whereas service charges are usually made to the persons receiving the service.

6. *Miscellaneous Revenues*—These may include proceeds from fees for private collection licenses, fees for salvage privileges, sale of salvaged material, or sale of collection privileges.

Since refuse collection represents such a large portion of any community operations budget, approximately 10 percent, it is important that both financing

---

[p]Numerous experts feel that the rising costs, developed technological capability, and importance of the waste management function, warrants a conversion of the collection and disposal of waste to the form of a regulated utility; see Reference 55.

methods and accounting procedures be carefully considered and monitored. The most recent accrual of data regarding methods of financing—as prepared by the American Public Works Association—is shown in Tables 4-3 and 4-4. These tables indicate finance methods in various cities surveyed using differing collection methods and of varying sizes. An effective accounting system can assist management to improve weak portions of an operation and can provide the needed exhibits when soliciting bank loans for new equipment, citizen support for bonds or taxes, increases in rates, and changes in collection frequency or methods. [56, 57]

Collection Costs

We have said a great deal about the high cost of the labor intensive collection function, the responsibilities for completing the task, and the methods that may be employed to finance and contract for the waste removal system implementation. But specifically, what are these costs? In the previously cited Reference 21, reporting is offered of recently acquired, specific cost data. A sample of 166 cities reported an annual total of 12,352,319 tons of solid waste collected at a cost of $217,040,288 for collection, or an average of $17.66 per ton. The average collection cost per ton is $9.50 for cities with populations of less than 100,000, $10.20 per ton for cities with population between 100,000 and 500,000, and $24.05 per ton for cities in the 500,000 and over population range. The last figure, however, reflects the weighting effect of both the huge tonnages and high collection costs of one or two large communities. For instance, New York City must collect once a day in crowded ghetto areas. In order to secure more typical cost figures, the *median* city in each population range has been determined. The figures for the smaller and medium sized cities remain about the same at $9.90 and $10.64 per ton, respectively; however, the median city in the largest population category has a collection cost per ton of only $12.78 as compared with the above mentioned weighted figure of $24.05.

There were large variations in reported collection costs per ton among the cities in every population category. Calculations based on the information submitted range from $1.56 to $80.00 per ton.[q] Some of the more extreme variations are obviously the result of inadequate records, or simple accounting

---

[q]Note that the collection costs in urban areas reflect a regularity attributed to predictable distribution of refuse along routes. In rural and public recreation areas, waste distribution and costs for collection range from $28 to $302 per ton. See, "Recreation Area Pickup and Disposal", *Solid Wastes Management/Refuse Removal Journal,* September 1971, p. 18.

Table 4–3
Methods of Financing Refuse Collection Services, by Size of Community, 1964

Population Size of Community	Total No. of Communities in Sample	Percent	Distribution of Financing Methods: General Tax	Service Charge	Tax and Service Charge	Other
5,000–9,999	180	100.0	47.2%	39.0%	13.4%	0.6%
10,000–24,999	307	100.0	46.0	38.0	16.0	0.0
25,000–49,999	190	100.0	51.5	32.7	14.2	1.6
50,000–99,999	93	100.0	58.0	28.0	12.9	1.1
100,000–999,999	74	100.0	59.5	27.0	13.5	0.0
1,000,000 and over	6	100.0	66.6	0.0	33.4	0.0
Total Sample	850	100.0	50.1	34.9	14.4	0.6

Table 4–4
Methods Used to Finance Refuse Collection by 857 Cities in 1964

Method of Financing	Method of Collection: M	M & C	M & P	M & C & P	C	C & P	Total	%
General Tax	209	25	81	20	68	26	429	50
Service Charge	149	24	37	4	74	11	299	35
Tax and Service Charge	69	10	29	4	8	4	124	14
Other	2	1	0	1	1	0	5	1
Totals	429	60	147	29	151	41	857	
%	50	7	17	4	17	5		

errors. Ignoring these extremes, however, the figures still indicate that wide cost variations are the rule rather than the exception. It is not uncommon for the refuse collection budgets of two cities in the same state with similar economics, levels of service, and populations to vary by 200 percent or more. This indicates great differences in the cost benefits of different collection systems.

In a sample of 39 cities, 29 cities reported a total average collection cost per ton of $9.52 for curbside collection in contrast with an average cost of $13.08 for 10 cities providing yard collection. Much of this differential could be attributed to differing cost accounting systems used by various municipalities.

The increased time and labor costs for carryout service reflect an approximate 37 percent higher collection cost per ton. These foregoing data are summarized in Tables 4-5 and 4-6. From this same sample of 166 cities and towns, a charting of frequency of collection is shown in Table 4-7 for 112 selected communities.

Collection Equipment

To understand the full working of transportation concerns, it is necessary to know about the costs and services performed by the numerous equipments available and utilized. It has been said that the best thing that ever happened to refuse collection was the mobile compaction truck. These commonly seen vehicles, which can carry three to four times the weight of a noncompacting truck of the same size, can reduce hauling times and numbers of trips needed—effecting enormous labor cost savings.[58] Mobile packers come in a variety of sizes and arrangements. They cost from $10,000 to $100,000 each; a thirty cubic yard unit with automatic transmission, however, usually ranges from $32,000 to $45,000. Use of mobile packers is increasing rapidly;[59] nonetheless, in 1968 only 97,000 mobile packers were in use in the United States compared with 179,000 of the nonpacker variety.

Based upon survey data, rear-loading mobile packer equipment receives the greatest use in refuse collection systems. Examples of two types of mobile packers are illustrated in Figures 4-3 and 4-4. An analysis of 5018 units of collection equipment has revealed that the four leading types in descending order of preference are rear-loading; side-loading; container; and front bucket. However, the last two types together comprise less than 5 percent of the total equipment used, while side-loaders comprise 8.2 percent. More than 87 percent of the units on the streets are rear-loading equipment. A number of cities, however, while using them for the major proportion of their collection activities, also report the need for auxiliary types of equipment for special functions

Table 4–5
Annual Solid Waste Tonnage and Collection Costs (166 Cities)

Population (1,000's)	*Tons (per year)*	*Collection Cost ($ per year)*	*Average Cost per Ton ($)*	*Cost per Ton for Median City ($)*
10 - 100	2,813, 819	26,757,188	9.50	9.90
100 - 500	2,803,700	28,605,200	10.20	10.64
500 and over	6,734,800	161,677,900	24.05	12.78
Total	12,352,319	217,040,288	17.66	

Source: Reference 21.

Table 4–6
Average Annual Cost per Ton Combined Averages (39 Cities)

	Curbside Pickup		*Backyard Pickup*	
Population (1,000's)	*Average Cost per Ton ($)*	*No. of Cities*	*Average Cost per Ton ($)*	*No. of Cities*
10 - 100	8.61	21	10.71	5
100 - 500	8.92	6	15.78	4
500 and over	20.71	2	14.09	1
Total	9.52	29	13.08	10

Source: Reference 21.

or unusual situations, such as spring cleanup or access problems in unusually narrow or winding roads. A number of cities use side-loading equipment almost exclusively. The preference for rear-loading equipment is most noticeable in large cities with populations greater than half a million. These metropolitan areas report use of 3106 rear-loaders compared with only 96 side-loaders; thus the former type of equipment comprises 94.5 percent and the latter 3.0 percent of the total units surveyed in this large-city category.

In cities of the 10,000 to 100,000 population size, these mobile packers in use have an average capacity of 18.5 cubic yards when used for twice weekly

Table 4–7
Frequency of Collection Service by Size of City (112 Cities)

Population (1,000's)	*Once Per Week*		*Twice Per Week*		*Three Times Per Week*	
	No. of Cities	*Total Population*	*No. of Cities*	*Total Population*	*No. of Cities*	*Total Population*
10 – 100	33	1,422,217	37	1,499,197	2	79,000
100 – 500	10	2,057,000	11	2,196,893	3	452,000
500 and over	8	5,397,180	7	4,564,393	1	114,000
Total	51	8,876,397	55	8,260,483	6	645,000

Source: Reference 21.

Figure 4–3. Rear-Loading Mobile Packer Courtesy of The Heil Co.

Figure 4–4. Front-Loading Mobile Packer Courtesy of The Heil Co.

collections. For cities in the 100,000 to 500,000 population range, the average capacity of vehicles is 17.8 cubic yards for twice weekly collection use and on once a week collection routes average capacity is 21.4 cubic yards. These statistics are clearly indicative of a need to evaluate equipment purchase practices [60] to achieve economies in collection operations.[61, 62, 63]

Of the existing American-made refuse collection equipment, the side-loading, packer type vehicle is probably best suited for one-man collection operations regardless of methodology. This equipment enables the operator to complete the collection task with a minimum of lost-time effort. With curbside collections, the side-loading packer equipped with right-hand drive may prove even more efficient. The costs involved in installing the right-hand drive equipment on the truck must be known in order to complete a cost-benefit study. Assuming a useful truck life of 5 to 8 years, if the crew completes 200 to 400 collection stops each day, even minor time savings can become significant over the life of the vehicle.

Rear-loading packers are satisfactory for one-man operation, although somewhat less efficient than side-loaders in terms of crew time. There is some disagreement within the industry on whether the rear-loading packer is more efficient for processing and compressing the refuse than the side-loading packer. The side-loading packer is more susceptible to the wind blowing light refuse materials out of the hopper; in addition, the packing mechanism tends to become less efficient as the full load capacity of the truck is approached. In most side- and rear-loading packers, a cycle time is involved, and lost time results when loaders are required to wait for the packing mechanism to complete its cycle to loading additional refuse.[r]

Containers used with mobile packers vary in sizes, shapes, and styles, depending upon the type of packer they are to be used with, the manner of dumping, and the manufacturer. The three general classifications of mobile packer trucks, front-loader, rear-loader, and side-loader, require containers of similar characteristics and these are identified by the same descriptive nomenclature. Containers having from one to six yard capacities are in general use. A container for use with a front-loader is shown in Figure 4–5. Some special industrial styles are somewhat shallow-pan shaped and may hold up to 12 or 15 cubic yards. The mobile packers which handle and dump these containers are equipped with special hoists and their several methods of use are described in detail below.

This class of container is made by many different manufacturers and used

[r]Side-loading mobile packers are available that permit continuous loading into the hopper.

Figure 4-5. Container for Use with Front-Loader Courtesy of The Heil Co.

with a variety of front-loading packers. The descriptions and comments that follow are general in nature and represent a composite picture of available equipment. Generally rectangular in shape and holding from one to ten cubic yards, these containers are intended for the temporary storage of loose wastes. They may be chute loaded under certain conditions but are usually filled by hand or from hand pushed carts. They are commonly placed on the ground and left in such a position that packer trucks may be easily maneuvered into position for the pick-up of the container by the trucks' loading mechanism. They are used extensively as parts of store, shopping center, apartment, and institutional waste collecting systems. The pick-up and emptying operations are handled by municipal or private hauling contractors' crews. After emptying into the mobile packer, the container is replaced on the ground for reuse.

Containers are equipped with hinged covers to prevent deposited refuse from being scattered by the wind. Some covers are spring loaded. The smaller sizes of containers are usually castered to allow for hand pushing. This is especially so where they are for use inside buildings where headroom and/or other clearances will not allow trucks to reach the usual locations of the containers. Where loading docks are available, the larger sizes of containers are feasible. They can be placed on the ground and loading from hand pushed carts can be accomplished from the dock level.

The designs of loading lugs and container shapes vary with the different manufacturers. In general, the lifting mechanism is a forked arrangement and requires matching slots or holders on the sides of the containers.

Front-loader packers, and hence the containers used with them, have the advantage of requiring less handling labor, since the truck driver has a much better view of the container to be lifted than he does with rear-loading packers. The front-loaders are usually operated by the driver, but in congested areas where maneuverability and vision are limited, a helper is required. Rear-loaders require a minimum of two men and commonly three make up a crew.

Prices of these containers will vary, depending upon design, weight, and corresponding delivery costs. However, they will generally be found in the following range: 1 cubic yard at $150; 2 cubic yards at $195; 3 cubic yards at $245; 4 cubic yards at $275; 5 cubic yards at $320; 6 cubic yards at $345; and 8 cubic yards at $415.

There are fewer types of rear-loaders designed to handle containers than there are front-loaders and, hence, there are a limited number of sizes and styles of rear-loading packer containers available. Shapes vary from generally rectangular, with sloping fronts for smaller containers, to large, somewhat shallow pan styles on the 10–15 yard sizes. Like the front-loader containers, these are equipped with hinged tops, some being spring loaded. The smaller sizes are usually caster mounted.

The side-loading packer container serves in much the same manner as does the front-loader container. Two general styles are common. One has a flat top while the other is slightly peaked. Both are equipped with lids. Sizes range from 1½ to 4 cubic yard capacities. Containers are castered and the manufacturer claims ease of handling and spotting. Weights of empty containers range from about 300 to 600 pounds.

Rear-loading containers differ greatly from those used with mobile packers and are not to be confused with them. These units are large special purpose containers, frequently used in industry. Open topped, tank type, and other styles of closed containers are available. They are equipped with special lifting ears or lugs and can be handled only by the special hoists. The rear-loading container is lifted by the special hoist, which is usually mounted on a short wheel, heavy-duty truck chassis. It can then be transported and subsequently deposited or dumped at a disposal site. Containers are generally of two basic types—tilt and skip, or bottom dump. Figure 4–6 shows a tilt frame hoist packer. In addition to styles already mentioned, hopper sludge, and pallet styles are available. Although, as previously mentioned, these containers are used in industry for hauling bulk

Figure 4-6. Tilt Frame Hoist Packer Courtesy of Tri-Pak, Inc., Louisville, Kentucky.

materials, they have applications for moving wastes around building complexes and can be chute or mechanically loaded.

Rear-loading containers come in several sizes ranging from one to ten cubic yards. The four most popular sizes and their approximate costs are:[S]

1.	Two cubic yards	$185
2.	Four cubic yards	$325
3.	Six cubic yards	$400
4.	Eight cubic yards	$500

Satellite collection vehicles are being used in some waste collection systems

[S]"Solid Waste Management in Residential Complexes", Greenleaf/Telesca, Planners, Engineers and Architects, Environmental Protection Agency Report SW-35c, 1971, Appendix F, p. A-70 – A-105.

to improve the efficiency of the pick-up crews.[t] These vehicles are small, motor driven devices, such as the example shown in Figure 4-7. They are commonly of the three wheeled scooter type, capable of handling the waste equivalent of a few barrels of trash. Such auxiliary vehicles are especially useful in residential neighborhoods (see Table 4-8), where the barrels or cans are placed at the end of the driveway near the house, or at the back of an alley. The satellite vehicle can be easily maneuvered on the driveway. Contents of the can are dumped onto the vehicle's bed and it is driven to the street where the large mobile packer truck is waiting. One trip per residence is usually sufficient for such an operation, and the time consumed in returning the can is eliminated. The satellite vehicle may incorporate a motor driven, dump bed to transfer the trash to the hopper of the larger truck. [64]

Collection System Innovations

Among the numerous innovations mentioned here regarding equipment and their efficient utilization—large mobile packers, [65] changing the frequency of pick-up using oversized mobile packers, [66] and more efficient scheduling of available packers [67] —two other very successful and innovative advancements are demonstrated by collection systems now used in Scottsdale, Arizona and Wichita Falls, Texas. These, respectively, regard (1) major equipment modifications and improvements and (2) route scheduling efficiencies promoted by use of a minitrain.

Scottsdale, Arizona is a resort city of 70,000 population, pressed for tax funds to provide municipal services. In 1965, the new city manager encouraged city employees to tinker on the job with new ideas for providing better service. Once a month the city manager sat down in his office with a group of employees, chosen by lot from their departments, to talk about their ideas. Those ideas that sounded promising were staffed out and the city manager and department head followed through. Bonuses were not paid. The only reward for a good idea was the satisfaction of seeing action taken. [68]

The city realized that refuse collection costs could be cut substantially with improved collection equipments—better containerized wastes and better mechanical machinery to dump the containers and pack the refuse. The city set out to develop this relatively simple concept with $100,000 of city funds, a

---

[t] Crew efficiency and collection cost benefits may or may not be improved through use of satellite vehicles. Data tabulated below is from "Satellite Vehicle Waste Collection Systems", U.S. Environmental Protection Agency report SW-82ts.1, 1972.

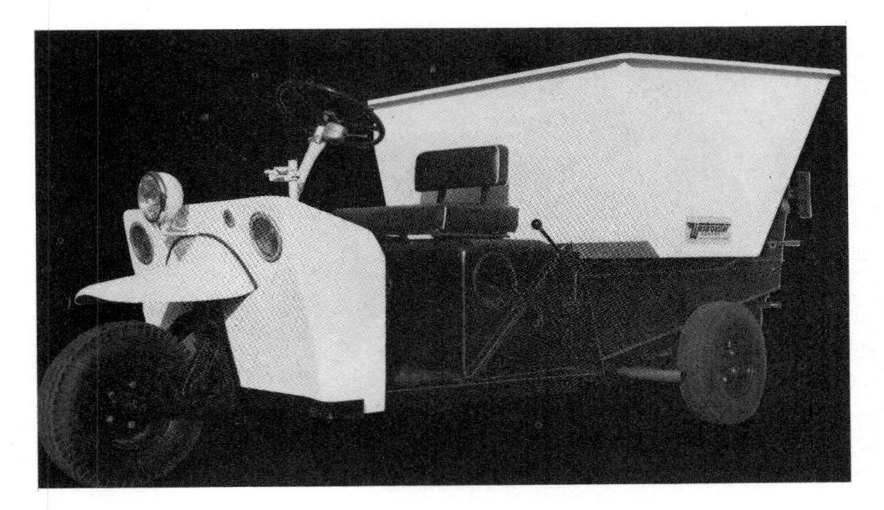

Figure 4–7. Satellite Collection Vehicle.

matching federal grant, and, ultimately, engineering assistance from industry. Large standardized plastic containers were manufactured and a modified front-end mobile packer was adapted to pick-up and dump the containers. The prototype vehicle shown in Figure 4–8 became known as "Godzilla." Ultimately an articulated mechanical arm was developed for curbside and alleyway pick-up of standard containers or bagged refuse.[69] The clamp at the end of the arm can adjust to handle either single-family 80-gallon plastic containers set out at the curb or four-family 300-gallon plastic containers designed for use on alley routes. In one hour, the truck and a single operator can collect the refuse of more than 100 residences at the curb or 350 in the alley. One such truck can collect twice a week from curb routes serving about 7,500 people or alley routes serving 15,000 people. The arm, which can reach over the hood of a parked car to remove a bag or container from the curb, cuts manual handling with its attendant costs and hazards. Since one man can drive the truck and operate the arm from the cab, labor costs are sharply reduced.[70]

The city supplies residents with 80-gallon, wheeled containers for individual curb pick-ups, or 300-gallon stationary containers placed in alleys for the joint use of four families. Collections are made by a 35-cubic-yard compactor truck equipped with a hydraulically-operated, telescoping arm which can grasp, lift, dump and replace containers in a complete cycle time of 18 seconds. The truck's driver operates the mechanism with electrical controls. He need never leave the truck cab and he has no helpers.[71] The truck, named the Barrel

Table 4–8
Residential Waste Collection Costs and Crew Efficiencies–Satellite Vehicle System Versus Conventional Methods

		Crew Efficiency (dwelling units per hour)		*Average Annual Cost/Dwelling Unit*	
Study Site	*Collection Frequency (per week)*	*Satellite Vehicle*	*Conventional (estimate)*	*Satellite Vehicle*	*Conventional (estimate)*
Atlanta, Georgia					
(Trash Taxi)	2	56	47	$33.50	$33.50
(Trashmobile)	2	83	65	21.50	24.00
Columbia, South Carolina	2	48	67	31.50	18.50
Knoxville, Tennessee	1	77	62	11.00	12.00
Medford, Oregon	2	113	85	25.50	30.50
Pasadena, California					
(Hilly)	1	38	36	46.00	44.50
(Flat)	1	71	45	28.50	37.50
Waukesha County, Wisconsin	1	61	50	17.00	17.00
All sites–once-a-week collection		64	51	21.50	23.50
All sites–twice-a-week collection		77	64	28.00	26.00

Source: J. E. Delaney, *Satellite Vehicle Waste Collection Systems,* Summary report (SW-82ts.1), U.S. Environmental Protection Agency, 1972, p. 13.

Figure 4-8. The Modified Front-End Mobile Packer "Godzilla."

Snatcher, and nicknamed "Son of Godzilla," is protected by patents. It is shown in Figure 4-9. By mechanizing its entire collection system, Scottsdale is cutting operating costs by $4 per capita, or almost $300,000 a year, which is a third of its solid waste management budget. That annual saving will soon amortize the capital investment.

The efficiencies and cost saving that are enjoyed by Wichita Falls, Texas, residents are derived from use of a train of carts that wind through the streets and alleys of the city and are efficiently utilized as satellite collection receptacles for conveyance of refuse to a mother vehicle.[72] The Wichita Falls demonstration, costing $270,000 over a three-year period with one third of the cost borne by the city, is to determine how solid waste disposal services of a city could be improved by developing a comprehensive management model of a solid waste collection and disposal system utilizing the container-train system of collection. Design of the studies, overall project management, and systems analysis and model development studies are being carried out by a consulting firm. Data correlating generation rates of commercial, industrial, and municipal wastes with

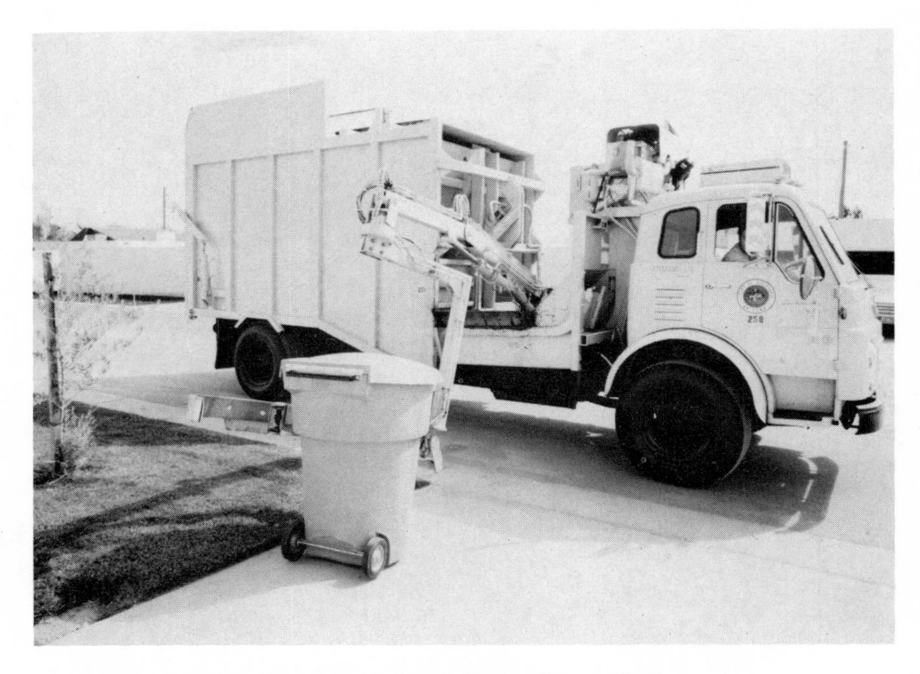

Figure 4-9. The Compactor Truck "Son of Godzilla."

land use have been developed. The annual fluctuation of solid waste generation has been determined. A detailed comparison of the existing packer collection system and the container-train system has been made. This includes documentation and evaluation of the economic, time and motion requirements; safety; reliability and productivity; aesthetics and public appeal; health aspects; and overall effectiveness of both systems. Optimization studies have been conducted to furnish information on how the collection operation should be performed to maximize the efficiency of the container-train system and minimize operational costs. These studies have varied routing of the trains, scheduling of the "mother" truck, and various other operational aspects of the collection procedure.

The entire systems analysis capabilities brought forth by computer technology should have a great impact upon waste management. Many administrative personnel in both the public and private sector have realized more efficiencies in collection through use of radio location of vehicles. However, this limited enhancement to the management function does not utilize the available technology of route modeling, synthesis of least cost alternatives, computerized billings, time and motion analyses, new fee structure generation, or optimization and tradeoff relationships for the men, machines, and management options available.[73-86]

The minitrain concept of train-transfer collection [87] has been utilized successfully in various locations; as applied in Tucson, Arizona;[88] as used in Winston-Salem, North Carolina, where 27 refuse trains are used to service six mother mobile packers;[89] as have serviced the community of Valdosta, Georgia for nearly a decade.[90] Heavy mobile packers have been replaced by ½-ton pick-up trucks pulling three containers mounted on pneumatic tires for efficient collection routings in Eau Gallie and Daytona Beach, Florida [91] and in identical means in Palm Springs, California.[92] The success of these innovative refuse collection systems is due to efficient utilization of manpower to fill a container train as it winds through an efficiently scheduled and routed pattern. The minitrain carries waste to a mother vehicle for compaction and removal, with a net saving to system operation of from 20 to 50 percent of labor cost, and negligible increase in capital equipment investments. A picture of the minitrain is shown in Figure 4-10.

Gulf Oil Chemical Co. has developed a Mechanical Bag Retriever (MBR) which is being used in Bellaire, Texas.[93] It is illustrated in Figure 4-11. Citizens on a volunteer basis dispose of their refuse in polyethylene bags, which are left out on the curb for pick-up. On collection day, the MBR is driven down the center of the street at a rate that equals or exceeds the pace of a three-

Figure 4-10. The Minitrain.

Figure 4-11. The Mechanical Bag Retriever.

Figure 4-12. The Litter Gulper.

man walking crew. The Retriever's hydraulic arm and pick-up basket are designed to gather the refuse bags from both sides of the street. The basket closes over as many as four bags at a time and the hydraulic arm automatically returns the bags to a hopper on the truck. The bags then move on a conveyor belt up into the body of the compactor truck. The entire process requires only one driver-operator, who never needs to walk or leave the cab of the truck. Application of this vehicle in many cities would be limited by local ordinances, street rights-of-way width, and traffic hazards.

American Can Company has sponsored the development of the Litter Gulper shown in Figure 4–12. This mechanical litter pick-up device was developed for routine cleaning of highway rights-of-way. The unit can travel at speeds up to 65 miles per hour when the arm and head are nested over the cab. When the unit is being used, the arm is controlled by the operator in the cab. Using rotating steel fingers, the Litter Gulper picks up bottles, cans, paper, and other litter while compressed air blows the items up the chute (arm) into an eight cubic yard compactor.

5 Pre-Disposal Concerns

In recalling the definition of the pre-disposal function, it is important to bear in mind that the intention of the function is to devise and implement methods for economy and efficiency and to utilize technology for improvement of policies and procedures in the solid waste management tasks. In actual daily practice this means better use of technology and management in the handling and processing of refuse prior to disposal. After refuse collection has been completed and refuse is transported away from the sites of generation and before the wastes are disposed through landfill or incineration, some important handling and processing may occur. These beneficial operations are discussed in this chapter—the pre-disposal activities that are of concern in modern solid waste management.

Of specific interest are the rail and barge hauling operations, the transfer station, the process applications of pulverizers and shredders, the use of balers, the tasks involved with segregation before disposal, and the management, economics, and personnel needs in the completion of pre-disposal functions. The most important items in handling aspects of pre-disposal are rail or barge haul and transfer stations which shall be addressed first.

One formidable problem that faces waste management personnel is the increasing transportation requirements for long distance cartage in removal of wastes from the collection point to suitable disposal sites. As trends toward urbanization continue to create megalopolis type cities, where land in almost any condition is at an economic and social premium, sites suitable for waste processing and land filling become more remote. To haul waste to these less accessible areas in small collection vehicles of the 20 cubic yard and 5 ton capacity type, creates high per ton disposal costs.[a] Large transfer trucks and trailers for refuse cartage with capacities of 60 to 120 cubic yards alleviate some of these high cost transportation problems, providing haul distances are not too great. A 60 cubic yard 15 ton refuse trailer can operate for about 35¢ per mile. The 60 cubic yard vehicle costs about $0.02 per ton mile while the smaller collection vehicle costs about $0.05 per ton mile for operating expenses.

[a] The operating cost of such vehicles as refuse compactor trucks runs about $0.25 per mile.

Therefore, the cost per ton mile for refuse cartage emerges as the significant cost factor for examination. Longer haul distances require trucks to be on the road for longer periods of time, and this in turn requires the purchase of more transfer trailer trucks and the hiring of more drivers to handle the same per day tonnage of waste.

Railhaul of Refuse

Railhaul of solid waste from large metropolitan areas is a reduced cost means of transportation compatible with very large and remote landfill facilities. Ideally, a landfill site used by such a waste system would be one of little value from the agricultural or development standpoint. Landfill sites are selected, in part, to increase the worth of the area when filled. The landfill must also have the potential for many years of use as waste repository. Rail cars have a decided advantage in that individual cars capable of holding 100 tons of waste can be used. When compared to the 15 ton capability of a 60 cubic yard transport trailer, the capacity of such a car and the car train, is impressive. Rail cars also offer an advantage in providing large volume temporary storage of wastes before cartage. Most importantly however, the cost per ton mile is under or approximately \$0.02.[b] While such transportation of solid waste has been hailed as a modern innovation that will assist in the cure of waste problems and help financially troubled railroads, it is not new.[c] A form of railhaul was used in Amsterdam, Holland, in 1878. Refuse was brought into the train yards in the outskirts of the city from outlying parts of the country, transferred to horse-drawn wagons, and carted to barges on nearby canals. The barges transported the waste to areas being reclaimed from the sea, where it was plowed into the soil. Today, railhaul is still used in Holland whereby refuse is transported to composting plants by specially designed large, top-loading, side-dumping rail cars.[2]

Today also, the concept of railhaul is receiving close scrutiny in this country.[3] Various aspects of railhaul have been studied and advocated by the rail industry in promotion of regional materials reprocessing centers served by waste trains,[4] in research regarding railhaul economics,[5] and for planning optimal routing of refuse trains between urban centers of generation and landfill

[b]Citing an example in Reference 1, it is estimated that Chicago can spend \$3 million annually to haul 1,300 tons per day by rail over 375 miles. This equates to \$0.0169 per ton mile for railhaul cost on a seven day per week operations basis, or \$0.0205 per ton mile for railhaul cost on a six day per week use factor.

[c]"Amsterdam Railhauled its Refuse a Century Ago", *Solid Wastes Management/Refuse Removal Journal,* July 1971, p. 11.

disposal sites.[6] These studies all indicate that railhaul systems must be reasonably large to gain economy of scale. Haul distances must usually be about 50 miles minimum to take advantage of the low ton-mile rate. Minimum tonnage must be about 1,000 tons daily, equivalent to a population of 200,000 to 250,000 if industrial wastes are included or to about 400,000 if they are not.[d] Additional studies, including those sponsored by state and local government and industry, endorse the railhaul concept when properly applied.[e] To maximize railhaul's economic advantages, shipments should utilize existing equipment fully. Loads of 100 tons per rail car are desirable. Schedules should fully utilize equipment by making 300 trips per year.

Because uncompacted solid waste from typical residential routes weighs only six pounds per cubic foot approximately, the number of rail cars needed to haul it becomes a function of bulk, not weight. To take advantage of the weight-carrying capacity of rail cars requires a density of 50 pounds per cubic foot. Refuse in a compactor truck weighs only 14 to 27 pounds per cubic foot depending upon type of refuse and compaction achieved. After being dumped, it expands, decreasing its density to about 12 pounds per cubic foot.

Researchers have discovered that high-pressure compaction at 2,000 to 3,000 psi produces stable high-density bales of residential and commercial solid wastes suitable for transporting by rail up to 700 miles.[7] Average density of the bales ranges from 60 to 80 pounds per cubic foot. Costs including straight depreciation, maintenance, and power, but excluding financing charges, return on investment, and labor are estimated at 40 cents per ton based on one eight-hour shift per day.

One firm is currently operating a high-pressure compaction station in St. Paul, Minnesota.[8] Refuse is dumped on the floor, shoved into a depressed conveyor, raised to a hopper, and squeezed into unstrapped bales measuring 3 x 3 x 4 feet. Each bale weighs about 2,600 pounds, a density of approximately

[d]R.R. Flemming, "Rail Haul", *Waste Age,* November-December 1971, p. 20–21.

[e]*Rail Transport of Solid Wastes,* Special Report No. 40, 150 p, is available from the American Public Works Association, price: $10.00; it reports the results of a thorough investigation of railhaul prospects. It was sponsored by 22 local governmental agencies, The Penn-Central Railroad, and the Environmental Protection Agency.

This report explores the feasibility of using railhaul as an integral part of a solid-waste management system. It includes five major studies:

1. Transfer stations and refuse processing–how to get the wastes into the system
2. Rail transport–how to get the wastes to the point of disposal
3. Disposal operations–how to dispose of large volumes of wastes
4. Administration–authority of states to establish a regional or area-wide authority which might operate a system
5. Public health and environmental control–how to overcome adverse environmental problems associated with a railhaul, solid-waste disposal system

70 pounds per cubic foot. At present, bales are trucked to a nearby landfill. However, this operation could be adapted to railhaul. Reclamation Systems, Inc., of Cambridge, Massachusetts, also had a high-density compactor in its disposal system but has since ceased operation.

Analyses indicate that railhaul projects should be conducted with unit trains. These are trains dedicated exclusively to hauling refuse. If freight cars full of refuse were attached to regular freight trains, delays would result. Cars would stand idle in yards while awaiting attachment. In contrast, unit trains would stand only while cars are loaded and unloaded. Assuming an initial solid waste shipment of one ton per capita per year, the standard metropolitan areas served by the Penn Central Railroad might, for example, supply more than 68 million tons of solid waste. This amounts to more than 1.3 million tons per week or 217,000 tons per day based on 312 working days per year. Stringent enforcement of environmental control regulations, which would close some marginal disposal facilities, could increase this tonnage by 10 to 15 percent. Using the existing rail network, it may be possible to solve for many years, about 70 to 80 percent of the solid waste-disposal problems in Illinois, Indiana, Ohio, Michigan, Pennsylvania, and New York.[f] Analyses suggest further that this might be accomplished by establishing only two to three statewide disposal sites per state. Finally, railhaul would greatly increase flexibility in the selection of disposal sites, because the increase in cost per mile of haul is very low once the system is established.

The railhaul of solid waste has been greeted with enthusiasm by some and disdain by others. Persons in areas from which the waste is removed by rail generally feel that this is a practical way of eliminating one of society's problems, which serves as an example of the "out of sight, out of mind" syndrome. Persons in areas where the railhauled waste is passing are less than enthusiastic. Those residing near the landfills have shown opposition to railhaul, indicating that no one wants to accept another person's trash. The problems of odor, contamination, and unsightliness are the most common objections given by dissenters. These could perhaps be overcome by milling the refuse, adding chemical deodorants, and vermin repellents. Business opportunities in waste railhaul are perceived by some persons living near railhaul terminal points, as they anticipate the time in the future when baled refuse in landfills may be economically mined for recoverable resources. According to one news item,[g] the

---

[f] Other railroads and cities are seriously considering the concept. Reference 9 describes a Southern Railway study for the Atlanta Region Metropolitan Planning Commission that would accommodate 1,000 tons fo refuse daily, process the wastes, and railhaul fifteen cars of baled materials to landfill each day.

[g] "Bring Us Your Cans, Your Coffee Grounds, Your Old Fish Heads," *The Wall Street Journal,* October 22, 1971, p. 1.

residents of Craigsville, Virginia, believe that if railhauled wastes from Washington, Baltimore, and Philadelphia were brought to their impoverished community, the trash could be processed with local labor to recover saleable iron oxide, compost, and fuel. In the proposed plan, the value received from these materials, coupled with fees charged the cities for acceptance of the trash, would assist in reviving the economics of the depressed community.

One of the more ambitious plans for railhaul and landfill has been proposed for use in the state of Washington.[h] The solid waste from the central and southeast portions of the state, as well as parts of Oregon, would be railhauled up to 90 miles and be deposited in old strip coal mines. Forecasts are that the 20,000 acre landfill at Centralia, Washington, could handle all the solid waste from the surrounding population of 2.6 million people for the next 35 years. The proposed waste management area to be served would include Seattle, Washington, as well as Vancouver, British Columbia, and Portland, Oregon and other smaller cities.

The refuse would be transported during the night in sealed rail cars, and the same cars could be used to carry construction gravel on their return trip. It is estimated that the hauling costs for the waste will be $5.50 per ton, a price competitive with existing disposal charges in Seattle. A 3,000 ton daily volume could be hauled initially, but this quantity would be increased to 5,000 tons per day in the future.

Western Pacific Railroad Company and San Francisco jointly studied the possibility of a 375-mile solid waste railhaul.[i] The trash can be transported to a high desert region in Lassen County in Northern California where it would be buried in a sanitary landfill. The railroad proposed the investment of $7 million to be spent on 150 new flatcars and 400 specially designed containers to hold the trash. Each car would have the capacity of 70 or more tons. Western Pacific believed that 1,300 tons of waste per day would be processed through such a system.

The Lassen County board of supervisors favored the plan and sought a 25¢-per-ton payment for rubbish accepted at the landfill. Another appealing factor to the county was the granted option for the railroad to buy 4,500 acres of desert land at $20 per acre. One major source of objection to the program was from residents of Las Plumas County, an area through which the trains would pass. The Mountain View landfill was an attractive alternative that was ultimately selected. Per ton costs for the system were estimated at $4.31. The

---

[h] "Proposal to Railhaul Wastes from 2.6 Million People, 180 Miles to 20,000 Acre Site", *Solid Wastes Management/Refuse Removal Journal,* May 1971, p. 54.

[i] "San Francisco Studies Shipping Trash by Rail 375 Miles for Disposal, *The Wall Street Journal,* November 12, 1968, p. 17.

possibility of salvaging steel scrap at the desert site and hauling it back to the bay area mills on the same trains was considered as a means of recouping some costs.

Other metropolitan centers have studied the advantages of railhaul of solid waste. In almost every case some railroad serving the city has been instrumental in examining and promoting the system.

Philadelphia has planned to utilize facilities of the Reading Railroad to remove wastes from the city for landfilling of abandoned strip mines in Schuylkill County.[j] The railroad would design and build the enclosed cars to haul 50 to 60 tons of waste; up to 1,200 tons of rubbish a day would be hauled at a cost of $5.74 per ton, or $2 less than the city incineration costs and exclusive of ash residue disposal costs.[k] The project includes treatment by shredding of refuse at one of two proposed transfer stations to be built by the city for $5-million. The Philadelphia railhaul project, initially announced in 1967, was later turned down by the city council.[l]

Denver and the Denver Rio Grand Western Railroad studied the possibility of railhauling that city's trash. The railroad claimed that removal of 875 tons of refuse daily could be achieved at a cost of $4.61 per ton.

Milwaukee tentatively approved a railhaul plan in 1969. This joint effort included the Chicago, Milwaukee, St. Paul, and Pacific Railroad, and the Acme Disposal Corporation of that city.[m] Charges for transporting 480,000 tons of waste per year would be $6.23 per ton. These would drop to $5.45 per ton after hauled tonnage reached 960,000 tons per year. The cost estimates were, once again, $2 per ton less than a proposed county-wide incineration system. Three, 1,000-ton per day transfer stations would augment the system.

Washington, D.C., railhauls some of its waste 25 miles to a landfill owned by the sanitation district.[n] The Richmond, Fredricksburg, and Potomac Railroad is used for transportation. The District of Columbia covers 65 square miles and serves more than 800,000 residents. The transfer station used in conjunction with the railhaul accommodates four gondola rail cars at a time as well as some transfer trailers. As antiquated incinerators are phased out of the Washington sanitation system, landfill, with rail, barge, and truck hauling is expected to gain increased use.

---

[j] "Railroad Will Haul Waste to Abandoned Strip Mine", *Solid Wastes Management/Refuse Removal Journal*, October 1967, p. 14.

[k] "Railroading Trash Out of Town", *Business Week*, July 6, 1968, p. 50.

[l] "Railroad Project for Philadelphia is Dead", *Solid Wastes Management/Refuse Removal Journal*, February 1968, p. 34

[m] "Milwaukee Picks Railhaul Instead of Incineration", *Solid Waste Management/Refuse Removal Journal*, April 1969, p. 10.

[n] "Capital City Using Railhaul Long Before Other Areas", *Solid Waste Management/Refuse Removal Journal*, May 1969, p. 18.

The use of existing subways have been suggested to augment rail transfer and as a place for loading of cars in dense urban areas. Utilization of this approach would necessitate late evening collection of refuse, and waste cartage at night, coupled with effective odor control in public-use facilities.

Barges can be used in a similar manner to railroads when navigable waterways are available.[10] Barges are built to about 550 ton capacity and are the most economical mode of bulk waste transportation when processing or disposal sites are more than 20 miles from the collection area. Barges are presently in use for transfer of refuse to New York Bay landfills such as at Fresh Kills. Barges can also be used for at-sea refuse disposal—provided legislation allows this continued use of the ocean environment. By high compression baling or slurrying wastes,[o] they may be made dense enough for sea disposal without refuse flotation. Open sea barges require sturdier construction than inland waterway vessels and construction costs run approximately 10 percent higher. New barge designs allow barges to be rolled over at sea for dumping of refuse bales and can be constructed for piggy-back carriage or railcars.

Table 5-1 summarizes the anticipated costs of the transport modes to the disposal function. Note that the compactor trucks used in these examples are 20 cubic yard vehicles with three man crews and that all per ton costs include preprocessing, use of transfer stations, transportation, and both vehicle loading and unloading costs.

Transfer Stations

As the haul distances increase, the per ton cost of refuse removal by truck goes up. This is especially true if multiman crews are used on each truck. The men on the crew, other than the driver, are idle while the truck makes the trip to the processing or disposal site. One method of cutting this idle time cost is to transfer the waste from the smaller, maneuverable, neighborhood collection vehicle to some convenient storage area. The material can be eventually hauled to remote sites by a larger conveyance, preferable with a single driver. This, the turn-around for the primary collection truck is decreased, and the hours spent on the road with idle crews as passengers are eliminated. This approach to

---

[o] Dr. Robert Erb of Franklin Institute in Philadelphia has stated that refuse may prove to be a marine environment enhancement because of nutrients contained in refuse. Dr. Melvin First of Harvard's School of Public Health has indicated that incinerator residue slurries of 15 percent by weight of ash have not had adverse effect upon shellfish or sea plants. Other experts have recommended consideration of deep ocean disposal of densely baled refuse beyond the continental shelf along and in riffs in the ocean floor's geologic plates.

Table 5-1
Comparison of Transport Modes Round Trip Costs in Dollars per Ton

	One Way Mileage to Site									
Mode	*10*	*20*	*30*	*40*	*50*	*60*	*70*	*80*	*90*	*100*
Compactor truck at 10 mph	7.25	–	–	–	–	–	–	–	–	–
Compactor truck at 20 mph	3.80	7.40	–	–	–	–	–	–	–	–
Compactor truck at 40 mph	2.20	4.40	6.60	–	–	–	–	–	–	–
17 Ton Semitrailer	1.95	2.60	3.20	3.65	4.10	4.55	5.00	5.40	5.85	6.25
Railhaul	–	–	4.00	4.10	4.15	4.25	4.35	4.45	4.55	4.65
Bargehaul	2.30	2.65	2.85	3.00	3.15	3.20	3.35	3.50	3.60	3.70

Source: Foster D. Snell, Inc., "Comprehensive Short-Range Solid Waste Study" for New York State Department of Environmental Conservation, Albany, New York, p. 21.

logistically prestaging of collected refuse at a central point for economical processing and transportation involves use of what is termed a transfer station.[11] This technique of management of the collection function has been rapidly gaining acceptance and wider usage in cities throughout the country.[12, 13]

These transfer stations take various forms. One type is a large building structure holding many truckloads of waste located conveniently near the residential or commercial area where refuse is generated. This building could contain heavy-duty compacting equipment, or it could be little more than an intermediate bulk storage building. There are some problems of social acceptance with such a transfer station. It is akin to an open dump in the eyes of some nearby residents. It increases heavy truck traffic in the area and if improperly constructed or operated, it will create objectionable odors and harbor flies and vermin. However, a well constructed and operated transfer station with appropriate landscaping will go far to solve these problems.

Another form of transfer station is the large tandem trailer truck body. It may consist of a trailer, mounted on wheels and axles, or large roll-off bodies or boxes that are later lifted onto a flat-bed truck body. There are some advantages to these portable transfer stations. They can be placed into service in the area where the demand is high, and then subsequently can be removed to another busy area. Some persons object to leaving these transfer stations or trailer bodies on the street and may object ot them becoming a permanent structure in the neighborhood.

Barges and railroad cars are also used as transfer stations where long haul to a disposal is expected.[14] Such carriers are usually parked in areas where the exposure of the waste is not as controversial as it would be in a residential or commercial neighborhood.

The cost savings possible with transfer stations or transfer trailers will depend upon unique circumstances of the particular solid waste management system where the concept is applied. A general comparison of the per ton costs between direct-haul and transfer stations has indicated [15] that the transfer concept becomes practical when round trip hauls are approximately eighteen miles.

Further, the point at which transfer stations become break-even operations is directly related to the type of truck performing the primary collection. It has been learned from the South Gate Transfer Station of the Sanitation District of the City of Los Angeles, that the smaller privately owned trucks, with higher dollar per ton per minute operating costs, are better able to take advantage of the transfer station than larger municipal vehicles. The larger truck, with lower dollar per ton per minute cost, can make longer hauls to the final site and still operate cheaper than if a transfer station is utilized.

The largest transfer station in the world[p] was opened for operation in late 1970 by private contractors near the border of San Francisco and San Mateo counties in California.[17] The facility cost $3 million, operates 24 hours a day, and can accommodate 110 collection trucks an hour or 5,000 tons of solid waste a day in its 17 loading bays. This private venture is jointly owned by Golden Gate Disposal Company and Sunset Scavenging Company who have formed a subsidiary–Solid Wastes Engineering and Transfer Systems–to manage the operation. These firms presently handle collection and disposal of 1,700 tons of refuse a day generated by 700,000 residents.

Incoming collection trucks are first weighed and a record is made of the owner, truck, and in-out time. These data are placed into a computerized system where they are periodically summarized for accounting purposes.[18] An incoming truck backs into one of the 17 bays and its load is dumped into the reinforced concrete pit. Transfer trailers are loaded from the pit. Dozers push the mixed refuse to the trailers, but large heavy waste items such as refrigerators, stoves, and water heaters are loaded by cranes. The crane operators also distribute the weight in the transfer vans by placement of these larger items. These cranes also compress the loose lightweight material to maximize the trailer loads.

The transfer vans–25 ton tractor trailers–haul the refuse to Mountain View landfill, some 32 miles south of San Francisco. Here, the trailers are backed onto hydraulically operated tipping ramps and raised to the dump position. The trailer can be tilted up to 70 degrees for gravity discharging of the contents. Dumping time is approximately six minutes.

The transfer station concept is not restricted to large metropolitan areas. San Bruno, California, a city of 30,000 in the San Francisco bay area, generates better than 100 tons of refuse a day. The nearest landfill is 40 miles away and the small collection trucks required to properly service the residential routes cannot economically make such a trip to empty their load.[q]

A transfer station has been constructed, and is operating quite satisfactorily, on a lot measuring only 50 by 100 feet. The small collection trucks discharge their waste into either one of the two stationary compactors that load 70 yard, 20 ton transfer trailers backed up to the structure. The trailers are held in place by steel guides imbedded in the concrete building. All waste remains enclosed in the building, compactor, and transfer trailers, preventing scattering. If the

---

[p] Each activity of the transfer operation is monitored by computer as described in Reference 16. Programming is to afford maximum facility flexibility to reclaim salvageable items.

[q] "Mini-Transfer Station Concept Fills Small Urban Needs", *Solid Wastes Management/ Refuse Removal Journal,* July, 1971, p. 19.

system is temporarily overloaded, or both trailers are on the road, refuse is deposited on the floor.

Pre-Disposal Processes and Equipment

As has been mentioned, processing equipment used at transfer stations is numerous.[19] Particularly useful are the compactors and balers utilized to compress refuse for greater economy in transport.[20] An additional benefit is derived from the compaction of wastes to in excess of 60 pounds per cubic foot, about the density of earth.[21] Compression sharply curtails decomposition rates through biological activity.[r] The baling process squeezes out virtually all of the oxygen in the garbage mix, starving existing bacteria and sharply curtailing future bacterial growth on the interior of the bale. It is the bacteria in unprocessed garbage that normally creates smell which attracts vermin and birds. The baler, as a type of compacting equipment, reduces the original loose volume of compressible materials fed into it, and produces a dense compact package of manageable size and weight. Some finished bales are bound with metal strapping, wire, or twine; others are retained in a plastic bag or a corrugated board box which may be bound by restraining strapping or wires. Transfer stations sized balers are machines capable of producing bales weighing from 150 to 1800 pounds or more and having volumes from about 6½ to 35½ cubic feet.

Balers do not lend themselves well to the handling of excessively wet wastes. Bagged or loose wastes can best be handled by this equipment. If salvage of wastes is to be achieved, then selective segregation must be practiced prior to baling. In municipalities utilizing sanitary landfill for disposing of solid wastes, the baling of refuse can provide a satisfactory processing method, but some cities require that baling wire be cut at the disposal site. Both advantages and disadvantages are presented and use of balers must accommodate further processing and disposal intentions. Primarily,

1. When refuse is compacted to such high densities, personnel need mechanical aids to transfer refuse from the baling equipment to the transport vehicle.

2. When the high-density refuse is charged into an incinerator, it remains almost intact. Conventional incinerators char the outer surfaces, but the internal volume is basically untouched, and volume reduction achieved through burning is very small.

---

[r]Studies described at length in Reference 22, indicate reduction in BOD (biochemical oxygen demand) by a ratio of 30 to 1 and COD (chemical oxygen demand) reduction by a 40 to 1 ratio in highly compacted refuse over loose wastes.

3. If the refuse is to be deposited in a landfill, the high density achieved is of definite advantage, since the effective landfill volume used is one-half that used after ultimate settling of raw refuse, under conventional disposal practices.

Other types of processing equipment find use in the pre-disposal function and include chippers, shredders, pulverizers, and grinders. Chipping of refuse is generally restricted to wood products, although other materials may also be processed to a small degree. The chipping units are useful in reducing the volume of demolition and construction materials, tree trunks or branches, wooden furniture, and bulky wastes. Chipping is rapid and efficient and results in small particle sizes which are highly acceptable for incineration or compacting. The high surface area to volume ratio of chips will usually result in highly efficient combustion and a minimum of ash residue. The low void volume in a pressed brick of chips permits more efficient utilization of landfill volumes.

Shredding is much like chipping, except, instead of "shaving" small particles off a refuse component, the pieces are torn from it in a ripping manner. [23] Shredding results in larger particles or pieces than chipping and, therefore, utilizes more landfill space when disposed of immediately. If the shredded pieces are to be incinerated, the resultant volume is basically equal to that achieved with chips. Shredding of automobile bodies has become a standard predisposal method for volume reduction. Shredders are made that routinely process 60 to 100 auto hulks per day.[s]

Dry pulverizing is a hammering or crushing technique which reduces the refuse material to a dry pulp of small particle sizes. [24] When fibrous materials are pulverized, they may exhibit some matting if strands are not completely broken, and separated. The pulverized material may also agglomerate if the particles are very small.[t] These materials form high density briquettes upon compaction [25] and when incinerated exhibit very high volume reductions. Becuase of the smaller particles produced by pulverizing, higher disposal densities in a landfill can be achieved and resultant landfill life is further extended.

Grinding has been attempted both on-site and at a central processing location. [26] Frequently the term grinding has been applied to other fragmenting processes such as shredding and pulverizing.[u] The grinding technique is one in which refuse is introduced between two hard surfaces and abraded. Grinding is

---

[s]B.W. Brown, "Big Shredder", *The American City,* July 1970, p. 70–71.

[t]Dry pulverization reduces waste volume by 50 percent and even more if the refuse is wetted after milling.

[u]A large refuse grinding station, installed in Chicago in late 1970, contains grinding equipment costing $127,000 in a $900,000 processing facility which includes other dry process machinery.

a dry process with only the included liquids wetting the product. The resultant product is a very fine particle "powder." The powder will tend to agglomerate due to moisture or particle electrostatic charges developed during the abrasion process. The process is slower than those previously discussed, but results in a product capable of higher density compaction than the others.[27] The incineration of such a powder is very complete, although difficult to control because of the particles' lightness. The combustion fire air will tend to suspend such particles, causing quicker clogging of air pollution controls. Pyrolyzation of powder requires no air. Incineration of a powder requires considerable less air than conventional refuse combustion because of the high surface area to volume ratio with powder. Heat transfer is very rapid. When compacted into briquettes for disposal at landfill sites, the volume required is less than other mechanical processes and increases landfill life proportionately.

Each of these dry processing techniques produces a milled product with certain advantages.[V] However, certain authorities dispute the claim that milling enhances effective incineration or landfill activities.[29] Nonetheless, refuse milling as a pre-disposal process has many supporters.[30]

Wet pulverization at large central processing plants is quite different than with residential building sized equipments.[W] Proponents of the process believe that the combination of pneumatic transport of refuse over short distances may be followed by wet pulverization and pipeline transit of refuse slurry economically over distances greater than 50 miles.[32] Use of this hydropulper concept suggests the collection of sewage and refuse as a single facility process.[33]

[V]As discussed in Reference 28, refuse milling seems to substantially reduce insect population at sites of waste storage and disposal.

[W]The pneumo-slurry system wet pulverizes untreated solid refuse and makes a 12 percent by weight suspended solids slurry that can be conveyed by pipeline.[31] Great volumes of wastes may be so processed for economic transport to disposal.

Bibliography

Chapter 2

Solid Waste Characteristics

1. "Comprehensive short-range solid wastes study.", Foster D. Snell, Inc., for New York City Department of Sanitation, Dec. 1969.
2. Sumichrast, Michael, "Profile of the builder and his industry.", National Association of Home Builders, Washington, D.C., 1970, 222 p.
3. Westerhoff, G. P., and R. M. Gruninger, "Population density vs. per capital solid waste production.", *Public Works,* 101(2):86–87, Feb. 1970.
4. Shin, K. C., and H. Straehle, "Influence of the structure and economy of an inhibited area on the amount and composition of domestic refuse and trade and industrial wastes.", *Wasser, Luft, und Betrieb,* 14(6):223–232, June 1970.
5. American Public Works Association, "Quantities and composition of refuse.", Municipal refuse disposal, 3d ed., Chicago, Public Administration Service, 1970, p. 21–55.
6. Burgess, R., "APWA Research Project 66–1: comparative public works statistics.", American Public Works Association Yearbook, Chicago, American Public Works Association, 1968, p. 190–207.
7. Golueke, C. C. "Chemical and microbial characteristics of urban solid wastes.", Presented at 1969 Annual Meeting American Society for Microbiology, Miami Beach, May 4-9, 1969. P. 1–17.
8. Michaelsen, G. S., and A. F. Iglar, "Disposing of disposables.", Presented at Annual Meeting American Hospital Association, Chicago, Aug. 21–24, 1967, Minneapolis, University of Minnesota, p. 1–8.
9. Hickman, H. L., "Characteristics of municipal solid wastes.", *Scrap Age,* 26(2):305–307, Feb. 1969.
10. Higginson, A. E., "The analysis of domestic refuse.", The Institute of Public Cleansing, 1956, 46 p.
11. "Methods of sampling and analysis of solid wastes.", Dubendorf, Swiss Federal Institute for Water Supply, Sewage Purification, and Water Pollution Control, 1970, 72p.
12. Schoenberger, R. J., and P. W. Purdom, "Sampling techniques for solid wastes.", Presented at National Conference on Industrial Solid Waste Management, University of Houston, Mar. 24–26, 1970, 14 p.
13. Ashby, R. Barry, "Experiment in New Haven.", *Waste Age,* 1:6, Nov.–Dec. 1970, p. 22–26.
14. "Collection, reduction and disposal of solid waste in multifamily highrise dwellings.", National Academy of Sciences, 1970.
15. Rao, S. S., N. Majumder, A. R. Ghosh, and S. K. Saha, "Character and quantity of refuse in rural homes.", *Journal of the Institution of Engineers,* 51(2):23–26, Oct. 1970.

16. Turner, J. M., "Residential solid waste production.", Solid waste studies, Gainsville, University of Florida, Aug. 1970, p. 44–47.
17. Galler, W. S., and L. J. Partridge, "Physical and chemical analysis of domestic municipal refuse from Raleigh, North Carolina.", *Compost Science,* 10(3):12–15, Autumn 1969.
18. Rogus, C. A., "Refuse collection and refuse characteristics.", *Public Works,* 97(3):96–99, Mar. 1966.
19. Outwater, E. B., "The disposal crisis. Our effluent society.", *National Review,* 22(7):203–204, Feb. 1970.
20. Niessen, W. R., and S. H. Chansky, "The nature of refuse.", Proceedings; 1970 National Incinerator Conference, Cincinnati, Ohio, May 17–20, 1970, New York, American Society of Mechanical Engineers, p. 1–24.
21. DeGeare, T. V., Jr., and J. E. Ongerth, "Empirical analysis of commercial solid waste generation.", *Journal of the Sanitary Engineering Division, Proceedings of the American Society of Civil Engineers,* 97(Sa6):843–850, Dec. 1971.
22. Burchinal, J. C., "A study of institutional solid wastes.", Morgantown, West Virginia University, 1968, 15 p.
23. "Per capita waste generation near thirty-five pounds a day.", *Chemical Engineering News,* 46(4):16, Jan. 22, 1968.
24. Page, W. W., "Solid wastes disposal in the chemical industry.", Proceedings; the Governor's conference on solid waste management, Hershey, Pennsylvania, October 8–9, 1968, Harrisburg Bureau of Housing and Environmental Control, p. 68–70.
25. Cummins, R. L., "A review of industrial solid wastes.", Bureau of Solid Waste Management, 1970, 41 p.
26. "The litter fact book.", New York, Glass Container Manufacturers Institute, Inc., 1960, 19 p.
27. "Physical characteristics of Rock Creek Park refuse.", 1966, 2 p.
28. Derrickson, C. F., "Motor vehicle abandonment in U.S. urban areas. Nature and extent of the problem, and adequacy of present methods of handling it.:, Washington, U.S. Department of Commerce Business and Defense Services Administration, Mar. 1967, 51 p.
29. Boyd, G. B., and M. B. Hawkins, "Methods of predicting solid wastes characteristics.", Washington, U.S. Government Printing Office, 1971, 28 p.
30. National Center for Resource Recovery, Inc., "The Municipal Solid Waste Stream: Its volume . . . its composition . . . its value", *NCRR Bulletin,* Spring, 1973.

Chapter 3

On-Site Concerns

1. Talty, J. T., "Are we being buried alive by solid wastes?", *Professional Engineer,* 39(2):47–49, Feb. 1969.

2. Trembly, J., "Times have changed since livestock were garbage disposal units.", Public Cleansing, p 141–145.
3. Linton, R., "In the Beginning.", *Terracide,* Little, Brown & Co., Chapter 8, p. 196–207.
4. Miner, J. R., "Raising livestock in the urban fringe." *Agricultural Engineering,* 51(12):702–703, Dec. 1970.
5. Black, R. J., "Safe and sanitary home refuse storage.", Public Health Service Publication No. 183, Washington, U.S. Government Printing Office, 1968, 6 p.
6. Brown, P., W. Wong, and I. Jelenfy, "A survey of the fly production from household refuse containers in the city of Salinas, California.", *California Vector Views,* 17(4):19–23, Apr. 1970.
7. Sabonjian, R., "How to make rat-proofing work.", *The American City,* 83(8):69–71, Aug. 1968.
8. Brown, R. Z., "Biological factors in domestic rodent control.", Rockville, Maryland, U.S. Public Health Service, 1969. 32 p.
9. "Paper sacks–a new refuse handling system.", National Refuse Sack Council, 1965, v.p.
10. Stone, R., "Survey reveals trends in the use of disposal refuse bags.", *Public Works,* 101(9):86–87, Sept. 1970.
11. "Refuse in the bag.", *Surveyor and Municipal Engineer,* 132(3987):30–32, Nov. 2, 1968.
12. McFeggan, J., "Bagged refuse brings better service.", *The American City,* 85(3):69–72, Mar. 1970.
13. Fox, G. G., "Paper-bag collection on request.", *The American City,* 70(10): 14, Oct. 1964.
14. Beck, A. H., "Impact of single-use refuse containers.", Proceedings; Fifth Annual Meeting of the Institute for Solid Wastes, Dallas, Sept. 29–Oct. 1, 1970, Chicago, American Public Works Association, p. 47–70.
15. "N. Y. garbage, a new market for treated paper?", *Chemical 26,* 5(10): 50, Oct. 1969.
16. Slatin, A., "Self-standing refuse bag, new horizon for polyethylene.", *Flexography,* 13(10):22, 58–59, Oct. 1968.
17. "Field study of disposable refuse bags in New York City; final report.", Ann Arbor, Mich., National Sanitation Foundation, Oct. 27, 1967, 6 p.
18. "New York City gives bags greatest test.", *Solid Wastes Management/Refuse Removal Journal,* 13(2):42, 44, 46, 48, Feb. 1970.
19. Sanborn, K. F., "Garbage cans? Who needs them?", *Public Works,* 96(3): 148, 150, Mar. 1965.
20. "Albuquerque changes system over to bags.", *Solid Wastes Management/Refuse Removal Journal,* 13(10):10–11, 44, 52, Oct. 1970.
21. "Argentina/Bagging garbage in PE.", *Modern Packaging,* 42(7):56, July 1969.
22. MacKay, D., "Challenge of a new town.", *Public Cleansing,* 57(5):261–270, May 1967.
23. "Paper sacks in Florence.", *Public Cleansing,* 58(3):134–135, Mar. 1968.

24. "80-Pound limit put on Los Angeles trash cans.", *Solid Wastes Management/ Refuse Removal Journal,* 12(21):24, Dec. 1969.
25. Appel. G. J., "Plastic trash carts to the rescue.", *The American City,* July, 1971, p. 77–78.
26. Messman, S. A., "An analysis of institutional solid wastes systems.", Urbana, University of Illinois, Jan. 1970, 61 p.
27. "Paper bags for waste disposal save hospital money.", *Modern Sanitation and Building Maintenance,* 17(2):19–21, Feb. 1965.
28. Wallace, L. P., "Solid waste generation by the units of a teaching hospital.", M.S. Thesis, West Virginia University, Morgantown, 1970, 99 p.
29. "Medics face giant problem in refuse handling.", *Solid Wastes Management/ Refuse Removal Journal,* 14(3):16, 46, 66, Mar. 1971.
30. "Universities survey central Philadelphia:, *Solid Wastes Management/Refuse Removal Journal,* 13(7):34, July 1970.
31. Spooner, C. S., "Solid waste management in recreational forest areas.", U.S. Department of Health, Education, and Welfare, Public Health Service, 1969, 133 p.
32. "New York considers using ads on cans, too.", *Solid Wastes Management/ Refuse Removal Journal,* 14(3):52, Mar. 1971.
33. Flintoff, F., and R. Milland, "Public Cleansing.", London, MacLaren and Sons, 1969, 475 p.
34. Bjoerkman, A. A., "Soviet streets are the cleanest.", *The American City,* 82(6):102–103, June 1967.
35. "Cleaning the littered highways—a costly job.", *Waste Trade Journal,* 66(26):9, June 27, 1970.
36. "Tin cans and jalopies add up to a costly headache for municipalities.", *Waste Trade Journal,* 66(22):13, May 30, 1970.
37. Cliburn, M. D., "Solid wastes from spectator events.", Solid waste studies, Gainsville, University of Florida, Aug. 1970, p. 11–12.
38. "Beach sanitation methods reviewed.", *Public Works,* 98(3):119–122, Mar. 1967.
39. Brown, J. R., and E. F. McDonough, "A model program for a regional system of collection and disposal of abandoned motor vehicles.", Hartford, Conn., University of Hartford, Regional Affairs Center, July 1969, 50 p.
40. Connolly, J. A., ed., "Abstracts; selected patents on refuse handling facilities for buildings.", Public Health Service Publication No. 1793, Washington, U.S. Government Printing Office, 1968, 320 p.
41. Bjoerkman, A. A., "Swedish underground pipeline vacuum network serves 3000 apartments.", *Solid Wastes Management/Refuse Removal Journal,* 10(3):12, 34, Mar. 1967.
42. "Waste transport by suction—a hygienic improvement.", *Stadtehygiene,* 19(12):4, Dec. 1968.
43. Orth, H., "Municipal cleaning service.", *Verein Deutscher Ingenieure Zeitschiift,* 112(10):646–650, May 1970.
44. "Governmental Refuse Collection and Disposal Association. Proceedings; Fifth annual seminar & equipment show, San Francisco, California, November 9–11, 1967.", 135 p.

45. Flintoff, F. "The collection of wastepaper.", *Public Cleansing,* 59(11): 578–582, Nov. 1969.
46. "The human factor. Does public understand?", (Atlanta II), *Solid Wastes Management/Refuse Removal Journal,* 13(10):14, 63, 65, Oct. 1970.
47. "Cans: where to place them.", *Solid Wastes Management/Refuse Removal Journal,* 13(3):26–40, Mar. 1970.
48. Cosby, W. A., "Refuse handling in high-rise, multifamily structures.", *APWA (American Public Works Association) Reporter,* 37(9):25–29, Sept. 1970.
49. "Waste disposal.", *Consumer Bulletin,* 53(11):38–40, Nov. 1970.
50. "Disposing of domestic waste biologically.", *Science Journal,* 4(2):21–22, Feb. 1968.
51. "Scrap system solves handling problem at Kraft Emballage.", *Boxboard Containers,* 76(12):69–71, July 1969.
52. Hopkins, T. F., "Stationary refuse compaction's expanding role in the handling of trash and refuse.", *Waste Age (Special Issue),* 22, 46, Sept. 1969.
53. Cummings, R., "The technology available for solid waste systems.", Proceedings; the Governor's conference on solid waste management, Hershey, Pennsylvania, October 8–9, 1968, Harrisburg, Bureau of Housing and Environmental Control, p. 85–90.
54. Kolb, J., "Garbage compaction is booming as alternative to incineration.", *Product Engineering,* 41(6):43–46, Mar. 16, 1970.
55. "Mobile Compaction, an innovation in waste disposal.", *Power,* 114(1): 164, Jan. 1970.
56. "A model contract for fixed packers.", 1971 Sanitation Industry Yearbook, New York, RRJ Publishing Corporation, 1970, p. 20, 54, 56.
57. McCall, J. H., "Utility concept and solid wastes systems.", Proceedings; Fourth Annual Meeting of the Institute for Solid Wastes, Cleveland, Sept. 16–18, 1969, Chicago, American Public Works Association. p. 1–7.

Chapter 4

Transport Concerns

1. Zausner, E. R., "Cost of residential solid waste collection.", Public Health Service Publication No. 2833, Washington, U.S. GPO, 1970, 24 p.
2. "The human factor—has it been neglected?", *Solid Wastes Management/ Refuse Removal Journal,* 18(9):38–39, 82, Sept. 1970.
3. "The human factor from the worker's angle.", *Solid Wastes Management/ Refuse Removal Journal,* 18(12):14, 32, Dec. 1970.
4. "The human factor, what should be done?", *Solid Wastes Management/ Refuse Removal Journal,* 14(1):30, 48, Jan. 1971.
5. Kazan, N., "Can free enterprize speed up or garbage collection?" *New York Magazine,* July 12, 1971.

6. "Five-year study of industry wage scale.", *Solid Wastes Management/Refuse Removal Journal,* 13(12):20–22, Dec. 1970.
7. "Hourly wage scales vary widely from east to west.", *Solid Wastes Management/Refuse Removal Journal,* 12(12):10, Dec. 1969.
8. "Is the cost justified?", *Public Cleansing,* 57(2):100–102, Feb. 1967.
9. Austin, H. H., W. M. McLellon, and J. C. Dyer., "Training the environmental technician.", *American Journal of Public Health and the Nation's Health,* 60(12):2314–2320, Dec. 1970.
10. National Safety Council. "Refuse collection in municipalities.", *National Safety News,* 99(4):33–40, Apr. 1969.
11. "State of California. Department of Industrial Relations. Disabling work injuries in refuse collection.", Work injuries in California, San Francisco, 1967, p. 3–9.
12. Van Beek, G., "Employee safety in the solid wastes industry.", *Public Works,* 99(12):74, Dec. 1968.
13. "Equipment danger markings.", *Solid Wastes Management/Refuse Removal Journal,* 13(7):6–7, July 1970.
14. Star, S., "Safety for refuse collection systems.", Proceedings; Fifth annual seminar and equipment show, San Francisco, California, Nov. 9–11, 1967, Government Refuse Collection and Disposal Association, p. 33–41.
15. Van Beek, G., "Personnel: accident prevention.", *National Safety News,* 99(4):41–44, Apr. 1969.
16. Hanks, T. G., "Solid waste/disease relationships, a literature survey.", Public Health Service Publication No. 999–UIH–6, Washington, U.S. Government Printing Office, 1967, 179 p.
17. Gellin, G. A., and M. R. Zavon, "Occupational dermatoses of solid waste workers.", *Archives of Environmental Health,* 20(4):510–515, Apr. 1970.
18. "Proposal for licensing refuse route/generation and simulation computer programs.", Baton Rouge, Owen and White, Inc., Consulting Engineers, 1969, 16 p.
19. Truitt, M. M., J. C. Liebman, C. W. Kruse, "Conclusions and summary.", Terminal report of an investigation of solid waste collection policies, 2 v., Baltimore, The Johns Hopkins University Department of Environmental Health, Aug. 1968, p. 135–152.
20. Wersan, S., J. Quon, and A. Charnes, "Mathematical modelling and computer simulation for the design of municipal refuse collection and haul service,", Final report, Northwestern University, Jan. 1970, v.p.
21. "A study of solid waste collection systems comparing one-man with multi-man crews; final report.", Ralph Stone and Company, Inc., Engineers, Public Health Service Publication No. 1892, Washington, U.S. GPO, 1969, 175 p.
22. "One-man collection crews get a boost.", *Nations Cities,* 8(3):26–28, Mar. 1970.
23. "100% paper-bag collection.", *The Amercian City,* 84(10):14, Oct. 1969.
24. Quon, J. E., R. R. Martens, and M. Tanaka, "Efficiency of refuse collection crews.", *Journal of the Sanitary Engineering Division, American Society of Civil Engineers,* 96(SA 2):437–453, Apr. 1970.

25. Bodner, R. M., E. A. Cassell, and P. J. Andros, "Optimal routing of refuse collection vehicles.", *Journal of the Sanitary Engineering Division, American Society of Civil Engineers,* 96(SA 4):898–904, Aug. 1970.
26. "Public works computer applications. Special Report No. 38.", Chicago, American Public Works Association, Aug. 1970, 143 p.
27. "Time and motion study in Atlanta aid to efficient route planning.", *Solid Wastes Management/Refuse Removal Journal,* 12(11):42–43, 66, Nov. 1969.
28. Stone, R., "How MTM can improve refuse collection.", *The American City,* Oct. 1971, p. 73, 76, 78.
29. Ralph Stone and Company, Inc., Engineers, "Field survey and analysis.", A study of improved refuse collection systems comparing one-man with multi-man crews, Los Angeles, June 1968, p. 3–44.
30. Rolfe, D. G., "Raising productivity in the refuse collection service.", *Public Cleansing,* 57(6):293–306, June 1968.
31. Cristofano, S. M., "One-man refuse collection crews.", *The American City,* 85(4):86–88, Apr. 1970.
32. Perl, M., "Principles of cost-saving in the use of garbage trucks.", International Research Group on Refuse Disposal (IRGRD), Information Bulletin 32, Rockville, Maryland, U.S. Department of HEW, Public Health Service, 1969, p. 17–21.
33. "Plastic liners now in use in Lebanon, Ohio, collection.", *Solid Wastes Management/Refuse Removal Journal,* 18(3):32, Mar. 1967.
34. Marks, D. M., and J. C. Liebman, "Mathematical analysis of solid waste collection.", Public Health Service Publication No. 2104, Washington, U.S. GPO, 1970, 196 p.
35. "Private Satellite Vehicle operators prove 22% faster than municipal units in six-area field test.", *Solid Wastes Management/Refuse Removal Journal,* Nov. 1971, p. 10.
36. "City-run collection services runs into financial problems.", *Solid Wastes Management/Refuse Removal Journal,* p. 16.
37. Clark, R. M., and B. P. Helms, "Fleet Selection for Solid Waste Collection Systems.", *Journal of the Sanitary Engineering Division Proceedings of the American Society of Civil Engineers,* Feb. 1972, p 71–78.
38. Schultz, G. P., "Managerial decision making in local government: facility planning for solid waste collection.", Cornell Dissertations in Planning, Ithica, Cornell University, Jan. 1968, 263 p.
39. Blair, L. H. and Harry P. Hatry, A. Don Vito, "Measuring the effectiveness of local government services–Solid waste collection.", Urban Institute, Washington, D.C. Oct. 1970.
40. "Private Haulers Debunk Claims Amateurs can do a better job.", *Solid Wastes Management/Refuse Removal Journal,* Apr. 1972, p. 8.
41. "New York drivers found cheating on load weights.", *Solid Wastes Management/Refuse Removal Journal,* 12(11):6, 52, Nov. 1969.
42. "Expands thousand acct Business by 700% in a dozen years.", *Solid Wastes Management/Refuse Removal Journal,* May 1972, p. 62.
43. "Solid Waste Removal is still a municipal responsibility.", Editorial in *The American City,* Apr. 72, p. 8.

44. "Place more reliance on free enterprise.", *Solid Wastes Management/Refuse Removal Journal,* 12(11):26, 36, 38, Nov. 1969.
45. Vanderveld, J., Jr., "Opportunities exist for public-private partnership.", *Solid Wastes Management/Refuse Removal Journal,* Nov. 71, p. 30.
46. Beerman, B., "City openly competes with private contractors.", *Solid Wastes Management/Refuse Removal Journal,* Nov. 1971, p. 8.
47. "Competition is keen in Finland's largest city.", *Solid Wastes Management/Refuse Removal Journal,* 13(3):8–9, Mar. 1970.
48. "Refuse collection in Arkansas.", *Public Works,* 99(6):158, June 1968.
49. Dysart, B. C., "An economic approach to regional industrial waste management.", Proceedings; 24th Industrial Waste Conference, Lafayette, Ind., May 6–8, 1969, Purdue University Engineering Extension Series No. 135, p. 880–895.
50. "Conversion to municipal refuse collection results in multiple benefits.", *Public Works,* 99(2):78–79, Feb. 1968.
51. "Baltimore marks first century of organized collection and disposal system.", *Solid Wastes Management/Refuse Removal Journal,* May 1972, p. 40.
52. "For haulers, Municipalities and counties–a sample franchise agreement.", RRJ Publishing Co., New York City, 1972 Sanitation Industry Yearbook, p. 12.
53. Clark, R. M. and R. O. Toftner, "Financing municipal solid waste management systems.", *Journal of the Sanitary Engineering Division, Proceedings of the American Society of Civil Engineers,* 96(SA 4):885–892, Aug. 1970.
54. Zausner, E. R., "Financing solid waste management in small communities.", Washington, U.S. Government Printing Office, 1971, 14 p.
55. "Solid waste is public utility, should be financed locally.", *Environmental Reporter,* Mar. 24, 1972, p. 1437–1438.
56. Zausner, E. R., "An accounting system for solid waste collection.", Public Health Service Publication No. 2033, Washington, U.S. GPO, 1970, 24 p.
57. Zausner, E. R., "An accounting system for solid waste management in small communities.", Public Health Publication No. 2035, Washington, U.S. GPO, 1971, 18 p.
58. Crawford, T., "An appraisal of compaction systems.", *Public Cleansing* 60(8):456–468, Aug. 1970.
59. "Firm trend to stationary compactors, greater truck size, diversity, more use of container trains.", *Solid Wastes Management/Refuse Removal Journal,* 11(1):82, Jan. 1968.
60. McClenahan, D. C., "Selection of equipment for refuse collection operations.", Proceedings; second annual meeting of the Institute for Solid wastes of the American Public Works Association, Boston, Massachusetts, Oct. 3–5, 1967, Chicago, p. 26–29.
61. Davies, A. G., "Why not mammoth collection vehicles?" *Public Cleansing,* 54(4):843–846, Apr. 1964.
62. Mendoza, E., "Larger trucks permit reduction in collection crew size.", *Public Works,* 99(4):106–109, Apr. 1968.

63. Griffin, T., "Refuse packer serves highway rest areas.", *Public Works,* 101(11):77–78, Nov. 1970.
64. Zapf, F. and H. Giles, "Mammouth trucks and mini scooters.", *The American City,* 32(1):77–79, Jan. 1967.
65. "A big 40-yard refuse truck.", *The American City,* 83(11):101–102, Nov. 1968.
66. "New mobile compaction system for bulky waste.", *Public Cleansing* 57(4):197–199, Apr. 1967.
67. Griffin, T. F., "Radio control of refuse trucks.", *Public Works* 97(2): 82–83, Feb. 1966.
68. "The struggle to bring technology to cities.", The Urban Institute, Washington, D.C., 1971, 80 p.
69. "Look ma, no hands!", *Heavy Duty Trucking,* 49(9):24–27, Sept. 1970.
70. Stragier, M. G., "Barrel Snatcher' eliminates refuse collection employment problems.", *Public Works,* 102(1):64–65, Jan. 1971.
71. Stragier, M. G., Director, "Mechanized residential refuse collection.", Department of public Works, city of Scottsdale, Arizona, Nov. 1970.
72. Roark, J. J., "Solid waste collection system design and operation—the Wichita Falls demonstration.", *APWA (American Public Works Association) Reporter,* 36(12):20–21, Dec. 1969.
73. Rao, S. A., "A multidisciplined approach towards regional planning and management of urban services.", Presented at First Pacific Regional Science Conference, East-West Center, Honolulu, Hawaii, Aug. 26–29, 1969, Berkeley, University of California, 1969, p. 1–11.
74. Betz, J. M., "Application of electronic data processing and operations research techniques.", Proceedings; Engineering Foundation Research Conference, Solid Waste Research and Development, University School, Milwaukee, Wisconsin, July 24–28, 1967, p. D-1.
75. Hume, N., "Sanitation management information system.", Proceedings; second annual meeting of the Institute for Solid Wastes of the American Public Works Association, Boston, Massachusetts, October 3–5, 1967, Chicago, p. 5–9.
76. Willard, W. B., "Systems analysis applied to solid waste management; the critical path method and program evaluation.", (Cookville) Tennessee Technological University, Nov. 1969, 23 p.
77. "Guidelines for local governments on solid waste management.", National Association of Counties Research Foundation, Public Health Service Publication No. 2084, Washington, U.S. Government Printing Office, 1971, 184 p.
78. Meresman, S. J., "PERT; concepts and application to solid waste management.", (Cincinnati), U.S. Department of Health, Education, and Welfare, 1970, 35 p.
79. Morse, N. and E. W. Roth, "Systems analysis of regional solid waste handling.", Public Health Service Publication No. 2065, Washington, U.S. Government Printing Office, 1970, 294 p.
80. "Mathematical modeling of solid waste collection policies.", Vol. 1 & 2,

Johns Hopkins U. under Grant No. UI-00539, GPO, D.C. PHS Publ. No. 2030.
81. Rothgeb, W. L., "Computerized refuse collection.", *Public Works,* 101(4): 88–89, 154, Apr. 1970.
82. Anderson, L. E., "Comprehensive studies of solid wastes management. A mathematical model for the optimization of a wastes maanagement system.", Berkeley, University of California, Feb. 1968, 62 p.
83. Austin, E., "Computer planning or refuse collection in Huddersfield.", *Public Cleansing,* 59(11):607–612, Nov. 1969.
84. "Design of optimal refuse storage depository system for servicing multiple dwelling unit buildings in economically disadvantaged areas.", Manchester, Conn., Meyers Electro/Cooling Products, Mar. 1971, 132 p.
85. Truitt, M. M., and J. C. Liebman, and C. W. Kruse, "Mathematical modeling of solid waste collection policies.", Vol. 1 and 2, Public Health Service Publication No. 2030, Washington, U.S. GPO, 1970, 311 p.
86. Helms, B. P. and R. M. Clark, "Locational models for solid waste management.", *Journal of the Urban Planning and Development Division, Proceedings of the American Society of Civil Engineers,* 97(UP 1):1–13, Apr. 1971.
87. Jackson, P., "Refuse container-train collection system results in savings.", *Public Works,* 98(2):74–75, Feb. 1967.
88. Danforth, H. L., "The train-transfer system of refuse collection.", American Public Works Association Yearbook, Chicago, American Public Association, 1963, p. 211–213.
89. Kildey, G. W., "Refuse train brings three-way savings.", *The American City,* 80(1):88–90, Jan. 1965.
90. "Collection by train.", *Public Cleansing,* 54(4):855, Apr. 1964.
91. "A 'train' system for residential collection.", Proceedings; Wyoming Seminar on Solid Waste Collection and Disposal, Laramie, Apr. 2–5, 1963, University of Wyoming, p. 52.
92. Essen, T. W., "Palm Springs reduces equipment costs using train system for solid waste collection.", *Western City,* 46(5):26, 28, May 1970.
93. Floyd, J., " 'Grab-bag' Texas style.", *The Orange Disc,* Gulf Oil Co., 19(12): 24–25, July-August 1971.

Chapter 5

Pre-Disposal Concerns

1. "Garbage ticketed for next fast freight.", *Engineering News Record,* 181(22):36, Nov. 18, 1968.
2. "Refuse transport by rail in Holland.", *Public Cleansing,* 57(2): 74–77, Feb. 1967.
3. Railway Systems and Management Association," The affluent and the effluent. Waste management–problems and prospects.", Chicago, 1968, 71 p.

4. Sheaffer, J. R., "Dimensions and trends in solid wastes.", Chicago, Railway Systems and Management Association, 1968, p. 35–38.
5. Wolf, K. W., "Rail haul as an integral part of waste disposal systems.", The affluent and the effluent, Waste management–problems and prospects, Chicago, Railway Systems and Management Association, 1968, p. 39–52.
6. Moore, E. J., S. Landy, M. O. Albl, D. H. Copenhagen, and D. M. Kerr, "The rail handling of municipal solid waste.", The affluent and the effluent, Waste management–problems and prospects, Chicago, Railway Systems and Management Association, 1968, p. 53–71.
7. Bugher, R. D., and K. Wolf., "Rail haul refuse disposal . . . using high-pressure compaction of wastes shows promising economics.", *The American City,* 83(8):79–80, Aug. 1968.
8. "The crisis in solid waste disposal.", St. Paul, Minnesota, American Hoist and Derrick Co., 6 p.
9. "Total system concept by Southern Railway system." Solid Waste Technical Memorandum No. 1", Atlanta Region Metropolitan Planning Commission, Apr. 30, 1970, 5 p.
10. "Transfer, shredding, rail, barge or vacuum chute.", *Solid Wastes Management/Refuse Removal Journal,* 13(9):6, 52, 54, 56, 104, Sept. 1970.
11. Stirrup, F. L., "Possible future development.", Transfer Loading Stations, London, The Institute of Public Cleansing, 1968, p. 27–40.
12. Nestner, M. L., "Introduction and historical review.", Transfer station feasibility study for City of Schenectady, New York, Troy, N. Y., Rensselaer Polytechnic Institute, Jan. 1969, p. 2–4.
13. Lausch, J., "How to transfer refuse–elegantly.", *The American City,* 83(10):85–87, Oct. 1968.
14. Evans, M., "Tranfer stations solve dump problems.", *Public Works* 99(5): 84–85, May 1968.
15. Gott, A. F., "Transfer stations beginning to dot countryside.", 1971 Sanitation Industry Yearbook, New York, RRJ Publishing Corporation, 1970, p. 16, 46, 94.
16. "Largest transfer point now in full operation.", *Solid Wastes Management/ Refuse Removal Journal,* 14(1):10–11, Jan. 1971.
17. "Largest transfer station opened by contractors.", *Solid Wastes Management/ Refuse Removal Journal,* 13(12):6, Dec. 1970.
18. Zausner, E. T., "An accounting system for transfer station operations.", Public Health Service Publication No. 2035, Washington, U.S. GPO, 1971, 20 p.
19. "Baling presses.", *Waste Trade World,* 110(21):13–22, May 27, 1967.
20. LaBarbara, J., "Solid waste compaction.", Proceedings; Institute on New Directions in Solid Wastes Processing 1970 (Forum), Framingham, Mass., May 12–13, 1970, Amherst, University of Massachusetts, Technical Guidance Center for Industrial Environmental Control, p. 152–162.
21. Menda, A., K. Kende, A. Insue, and M. Ito, "Decomposition of compressed mass of waste and water contamination.", Tokyo, Institute of Resources Research of Japan, 90 p.
22. Carver, P. T., "High density compaction processes for solid wastes.", Reuse

and recycle of wastes; Proceedings; Third Annual North Eastern Regional Antipollution Conference, University of Rhode Island, July 21–23, 1970, Stamford, Conn., Technomic Publishing Company, Inc., p. 198–224.

23. Johnson, D. S., "Don't burn it—chew it!, PW San Diego develops new disposal unit.", *Navy Civil Engineer,* 11(3):8–9, Mar. 1970.
24. Finnie, T. B., "Pulverization of solid waste.", *Public Cleansing,* 60(11): 606–609, Nov. 1970.
25. Keentop, D. C., "Pulverizing and compacting.", Proceedings; Institute of New Directions in Solid Wastes Processing 1970 (Forum), Framingham, Mass., May 12–13, 1970, Amherst, University of Massachusetts, Technical Guidance Center for Industrial Environmental Control, p. 146–151.
26. "Connecticut town orders grinders.", *Solid Wastes Management/Refuse Removal Journal,* 13(9):94, Sept. 1970.
27. Pikarsky, M., "Chicago looks to refuse grinding.", *Public Works,* 101(99): 82–83, Sept. 1970.
28. Gojmerac, W. L., E. C. Davenport, and D. W. Benjamin, "Effects of refuse milling of fly population.", *Public Works,* 101(6):88–89, June 1970.
29. "Pulverization—sense or nonsense?", *Public Cleansing,* 58(9):486–489, Sept. 1968.
30. "Poole plumps for pulverization.", *Public Cleansing,* 58(7):357–360, July, 1968.
31. Zandi, I., and J. Hayden, "A pneumo-slurry system of collecting and removing solid wastes.", Advances in solid-liquid flow in pipes and its application, Oxford, Pergamon Press, 1971, p. 237–253.
32. Zandi, I., "Collection and removal of municipal solid wastes by pneumo-slurry system.", *Compost Science,* 9(2):7–11, Summer 1968.
33. "One system treats sewage, solid wastes.", *Chemical and Engineering News,* 48(12):44–45, May 23, 1970.

Additional References

"A Cleaner Community," *Public Cleansing,* 60(9):491–494, Sept. 1970.

"A Daily Capacity 400 Tons Compression System is Adopted for Disposing of Bulky Waste in Osaka, Japan," Oct. 1970, 4 pages.

Aitken, I. M., and Robinson, G., "Refuse Collection Through Underground Mains," *Journal of the Society of Engineers,* 61(3);148–162, July–Sept. 1970.

Alexander, R. M. and Walters, J. W., "A Country-wide System—Chilton County, Alabama," *Proceedings,* Fifth Annual Meeting of the Institute for Solid Wastes, Dallas, Sept. 29–Oct. 1, 1970, Chicago, American Public Works Association, p. 1–13.

Althaus, G. F., "Continuous-Flow Refuse Collection," *The American City,* 81(12):90–92, Dec. 1966.

Bierbaum, P. J., Frand, W. R., Ruf, J. A., and Susag, R. H., "Composition of Solid Waste as Received at the Gainsville Compost Plant," Gainsville, University of Florida, 1969, 3 pages.

"Binational Solid Waste Management," U.S. Public Health Service, May 1970, v.p.

Boskoff, A., "Salvage Hauling Permit Scheme is Premature," *Solid Wastes Management/Refuse Removal Journal,* July 1971, p. 22.

Brewer, W. D., "Transportation Problems of Scrap," Remarks of W. D. Brewer, Commissioner of ICS, *Waste Trade Journal,* March 31, 1971, p. 11.

"Can We Reduce the Waste in Trash Disposal?", *Safety Maintenance,* 135(3): 43–44, March 1968.

"Change Day Collections to Nighttime Routes," *Solid Wastes Management/Refuse Removal Journal,* 13(2):14, Feb. 1970.

Clark, R. M. and Helms, B. P., "Decentralized Solid Waste Collection Facilities," *American Society of Civil Engineers, Journal of the Sanitary Engineers Division,* 95(5):1035–1043, Oct. 1970.

"Colorado, All Private Hauler, With Exception of 3 Cities," *Solid Wastes Management/Refuse Removal Journal,* March 1972, p. 34.

"Communicate with 26,000 Customers Through Newspaper Advertising," *Solid Wastes Management/Refuse Removal Journal,* 12(12):6–7, 22, 24, Dec. 1969.

"Contractors Handle Bulk of Pickups in Oregon," *Solid Wastes Management/ Refuse Removal Journal,* 13(7):13, 52, 54, 78, July 1970.

"Coping with the Garbage Crisis," *Design News,* 24(25):13, Dec. 8, 1969.

"Court Rules Trash Can is a Private Matter," *Solid Wastes Management/Refuse Removal Journal,* Sept. 1971, p. 8.

Culver, B. D. and Walsh, T. E., "Some Aspects of Systems Engineering for Waste Management," *Archives of Environmental Health,* 17(3):377–382, Sept. 1968.

Davidson, G. R., Jr., "A Study of Residential Solid Waste Generated in Low-Income Areas," U.S. Environmental Protection Agency, 1972, 14 pages.

"Drive Through Disposal Facility May be First in U.S.," *Solid Wastes Management/Refuse Removal Journal,* Perkins Township, Ohio, June 1971, pp. 11–12.

"Efficient Refuse Collection," *Public Works,* 99(4):136, Apr. 1968.

Engdahl, R. B., "Solid Waste Processing: A State-of-the-Art Report on Unit Operations and Processes," Public Health Service Publication No. 1865, Washington, D.C., U.S. Government Printing Office, 1969, 72 pages.

"Equipment and Ideas Improve Leaf Removal," *The American City,* 86(1):57, Jan. 1971.

Finkner, A. L., "National Study of the Composition of Roadside Litter," North Carolina, Research Triangle Park, Research Triangle Institute, Sept. 1969, 18 pages.

"Fiscal Year 1970 Outline of Cleaning Activities," Yokohama, Japan, Yokohamamashi Seisolyoku, 1970, 90 pages.

"For Solid Wastes: Squeeze and Package," *Chemical Week,* 102(14):55, April 16, 1968.

Fox, N. A. "Street Sweeping–Methods and Incentive," *Public Cleansing,* 58(2): 87–98, Feb. 1968.

Hart, C. J., "Collect Twice the Leaves with the Same Crew," *The American City,* 82(4):116–117, April 1967.

Haug, L. A. and Davidson, S., "Refuse Collection and Disposal Survey Indicates Changing Trends in 118 Western Cities," *Western City Magazine,* 40(4): 26, 28, 30, 31, 34, April 1964.

Hayden, J. A., Seidenstat, P., and Zandi, I., "Solid Waste Generation and Cost in Center City Philadelphia," Philadelphia, 1969, 30 pages.

Henningson, Durham and Richardson, Inc., "Survey of Solid Wastes," Collection and disposal of solid waste for the Des Moines Metropolitan Area. A system engineering approach to the overall problem of solid waste management. An Interim Report, Cincinnati, U.S. Department of Health, Education and Welfare, Public Health Services, 1968, p. 2(1–43).

Hickik, R., "How to Cut Street-Cleaning Costs," *The American City,* 79(7): 90–91, July 1964.

"How to Dispose of a Cooker," *Engineer,* 224(5849):353, March 1968.

"Industrial Services of America, Emphasis on Environmental Problems," *Secondary Raw Materials,* 8(8):112, 114, 115, August 1970.

Jaag, O., "A New International Reference Center," *APWA (American Public Works Association) Reporter,* 37(1):13, 14, Jan. 1970.

Jacobson, A. R., "Paper Bag Refuse System," *Public Works,* 97(4):146, 147, April 1966.

Jenkins, H. W., "The Monster," *The American City,* 85(9):146, Sept. 1970.

Joseph, J., "Fully Automated Refuse Truck," *Diesel and Gas Turbine Progress,* 36(12):18–19, Dec. 1970.

Katayama, T., "Solid Waste Transportation Through Pipelines," *Yosui to Haisu,* 11(8):631–639, Aug. 1, 1969.

"Keeping Employees Happy," *Solid Wastes Management/Refuse Removal Journal,* 13(2):8, 32, 36, Feb. 1970.

Kelley, B. C., "Solid Waste Generation and Disposal Practices by Commercial Airlines," *Solid Waste Studies,* Gainsville, University of Florida, August 1970, pp. 30–32.

Kutzschbauch, K., "Results of Studies Concerning the Bulk Weight of Domestic

Refuse in Berlin and its Compaction—1967–1968," *Zeitschrift fuer die Gesamte Hygiene and Ihre Grenzgebiete,* 16(8):593–597, August 1970.
Landman, W. J., "Designed to Work Hard," *The American City,* 79(3):106–107, March 1964.
Landor, W., "Packaging and 'The Third Pollution' ", *Packaging Design,* 10(3): 27–33, 62, May–June 1969.
"Litter Cleanup Now at $500 Million Annually," *Solid Wastes Management/ Refuse Removal Journal,* 13(10):6, 34, 58, Oct. 1970.
Markens, J. C., "Paper and Beer Cans Win," *The American City,* 86(1):53, Jan. 1970.
"Maryland Creates a Statewide Wholesale Sanitary District to Save Her Waterways," *APWA (American Public Works Association) Reporter,* 37(10): 21–22, Oct. 1970.
Mazowiecki, A. W., "A Big Truck and a Little Sweeper," *The American City,* 83(3):104–106, March 1968.
"Mechanization is Key to Low-Cost Tree Removal, *Public Works,* 99(1):104–105, Jan. 1968.
"Milwaukee County Settles for Baling and Burial," *Solid Wastes Management/ Refuse Removal Journal,* 13(1):6, Jan. 1970.
"New Collection Trucks Solve Sanitation Problem," *Public Works,* 99(8):109, August 1968.
"New Developments Built into Solid Waste Reduction Unit," *APWA (American Public Works Association) Reporter,* 37(9):30, Sept. 1970.
"On Site Mechanical Equipment for the Conditioning of Waste in Collective Dwellings," *Techniques et Sciences Municipales,* 64(5):168, May 1969.
Orndorff, R. L., "Regional Solid Waste Disposal Systems," Proceedings Second Annual Meeting of the Institute for Solid Wastes of the American Public Works Association, Boston, Mass., Oct. 3–5, 1967, Chicago, p. 9–25.
Owen, E. H., "Computer Program Cuts Cost of Urban Solid Waste Collection," *Public Works,* 101(1):60–62, Jan. 1970.
"Paper Sacks are Replacing Garbage Cans," *Waste Trade Journal,* 64(16):56, April 27, 1968.
Parkkivaara, J., "Russians Take Heed of Solid Wastes Problems," *Solid Wastes Management/Refuse Removal Journal,* 13(4):10, 42, March 1970.
"Pilot Plant to Separate Municipal Waste into Paper, Glass, Metal, Plastics, *Secondary Raw Materials,* 8(11):66, Nov. 1970.
"Proceedings; Ninth International Conference, International Association of Public Cleansing, Paris, June 1967, 277 pages.
"Proceedings; Wyoming Seminar on Solid Waste Collection and Disposal," Laramie, April 3–5, 1963, University of Wyoming, 77 pages.
Public Opinion Surveys, Inc., "Who Litters—and Why; Summary of Survey Findings Concerning Public Awareness and Concern about the Problem of Litter," New York, Keep America Beautiful, Inc., November 1968, 14 pages.
Quon, J. E., Tanaka, M., and Charnes, A., "Refuse Quantities and Frequency of Service," *ASCE Journal Sanitary Engineering Division,* 94(SA2):403–420, April 1968.

"Recreation Area Pickup Disposal Ranges from $28 to $302/Ton," *Solid Wastes Management/Refuse Removal Journal,* Nov. 1971, p. 10.

"Refuse in the Pipeline," *Surveyor Municipal Engineering,* 132(3982):1–2, Sept. 27, 1968.

"Refuse Industry in Bay City," *Solid Wastes Management/Refuse Removal Journal,* 10(5):28, 30, 32, 67, May 1967.

"Refuse Storage and Collection," Proceedings; 69th Annual Conference, Institute of Public Cleansing, Blackpool, England, June 1967, p. 1–8.

Rheinfrank, W. J., "Solving Street Cleaning's Toughest Problem," *The American City,* 79(4):94–95, April 1964.

"Re-using Solid Wastes," *All Clear,* 2(2):14, March-April 1970.

"Saigon Garbage Situation Serious Too," *Clean Air News,* 2(5):19–20, Feb. 20, 1968.

Shanholtz, M. T., "The Domestic Solid Waste Problem in Virginia," *Virginia Medical Monthly,* 96(11):670–671, Nov. 1969.

Sibiga, J., "Mathematical Methods Application to the Technique of Solid Refuse Removal," *Gaz, Woda I Technike Sanitarna,* 43(7):233–236, July 1969.

Sibiga, J., "Removal of Solid Waste via Pipelines,".*Gaz, Woda I Technike Sanitarna,* 44(7):239–243, 1970.

Singer, L., "Disposing of Disposables," *Design and Environment,* 1(3):25–29, Fall 1970.

Sliepcevich, E. M., "Effect of Work Conditions upon the Health of the Uniformed Sanitation Men of New York City," (Springfield, Mass.), Springfield College, June 1955, 253 pages.

Smith, D. C., "Storage and Collection of Household Refuse," American Public Works Association Yearbook, Chicago, American Public Works Association, 1963, p. 320–323.

"Solid Waste Collection and Disposal Operations," *Public Works,* 99(2):54, Feb. 1968.

"Solid Waste Management: A List of Available Literature," (Cincinnati), U.S. Environmental Protection Agency, April 1972, 35 pages.

"Solid Wastes Management Recommendations: A Plan for Immediate Action," Allegheny County, Pennsylvania, Solid Waste Advisory Committee, April 1969, 98 pages.

"Solid Waste Studies," Gainsville, University of Florida, August 1970, 50 pages.

Spradlin, B. C., "Solid Waste Management: A Systems Perspective," American Public Works Association Yearbook, Chicago, American Public Works Association, 1967, p. 181–184.

Stern, H. I., "Comprehensive Studies of Solid Wastes Management. Optimal Service Policies for Solid Waste Treatment Facilities, " Berkeley, University of California, May 1969, 85 pages.

Stragier, M. G., "We Automated Residential Refuse Collection," *The American City,* 85(11):66–67, Nov. 1970.

Tarbet, J., "Common Problems," *Public Cleansing,* 60(12):662–666, Dec. 1970.

Thatcher, R. N., "Optimal Single Channel Service Policies for Stochastic Arrivals," Berkeley, University of California, June 1968, 89 pages.

"The Heil Company," *Secondary Raw Materials,* 8(8):96, 98, August 1970.

"The Solid Waste Disposal Act as Amended," Title II of Public Law 89–272, 89th Congress, S. 306, October 20, 1965; Public Law 91–512, 91st Congress, H.R. 11833, October 26, 1970, Washington, D.C., U.S. Government Printing Office, 1971, 14 pages.

Toftner, R. O., "Developing a State Solid Waste Management Plan," Public Health Service Publication No. 2031, Washington, D.C., U.S. Government Printing Office, 1970, 50 pages.

Tucker, E., "New Garbage Service System Saves Manpower and Money for Glendale," *Western City,* 46(9):50, Sept. 1970.

"20-Yard Roll-Off Refuse Bodies," *The American City,* 84(8):119, August 1969.

"200,000 Customers to Please," *Solid Wastes Management/Refuse Removal Journal,* 13(5):78–80, 82, 92, May 1970.

Vaughan, R. D., "Solid Waste Management and the Packaging Industry," Proceedings; First National Conference on Packaging Wastes, San Francisco, Sept. 22–24, 1969, University of California, p. 115–119.

Walton, C. E., "Laramie's Experience with Garbage and Refuse Collection," Proceedings; Wyoming Seminar on Solid Waste Collecton and Disposal, Laramie, April 3–5, 1963, University of Wyoming, p. 30–34.

West, H. M., "Highway Litter," *Solid Waste Studies,* Gainsville, University of Florida, August 1970, p. 48–50.

"Western Pacific Railroad Company. Solid Waste Disposal Plan for the City of San Francisco," San Francisco, n.d., 16 pages.

Wheeler, J. G., "We Cut the Refuse Work Load . . . By a Change in Collection Schedules and Adopting New Collection and Disposal Methods," *The American City,* April 1972, p. 83.

Wilson, D. G., "Rethinking the Solid Waste Problem," *Science Journal,* 5(9): 69–75, Sept. 1969.

Zaltzman, R., "Refuse Collection–Tonnage Survey," Oklahoma City–County Health Department, Solid Waste Disposal Countrywide Study, Preliminary Report, Oklahoma City, C. H. Guernsey & Co., Consulting Engineers, 1967, p. 69.

Zepeda, F., "Comparative Study of Solid Waste Collection and Disposal Methods in the U.S.A. and Latin America: Solid Waste Term Problem," West Virginia University, Morgantown, Dec. 1968, 13 pages.

Index

Date Due

Due	Returned	Due	Returned
OCT 07 1987	SEP 18 1987		
APR 21 1988	APR 22 1988		
NOV 03 1988	NOV 08 1988		
JAN 09 1989	DEC 07 1988		
FEB 13 1990	FEB 13 1990		
APR 25 1990	APR 17 1990		
MAY 15 1990	APR 28 1990		
DEC 17 1990	DEC 12 1990		
JAN 05 1994	NOV 16 1993		
AUG 01 1994	JUL 13 1994		
DEC 09 1994	DEC 14 1994		
JUL 21 1995	JUL 21 1995		
JAN 04 1996	NOV 18 1995		
MAR 18 1996	MAR 23 1996		
APR 28 1996	APR 22 1996		
DEC 17 1996	DEC 05 1996		